HORSE S

How to Develop Your I

HORSE SENSE

How To Develop Your Horse's Intelligence

by

HENRY BLAKE

Trafalgar Square Publishing

First published in paperback in the United States of
America in 1994 by Trafalgar Square Publishing,
North Pomfret, Vermont 05053

Reprinted 1995
Reprinted 1997

Printed in Great Britain by
The Guernsey Press Co. Ltd, Guernsey, Channel Islands.

ISBN 0-943955-89-0

Library of Congress Catalog Card Number: 94-60008

Contents

1: *Beginnings of Horse Sense*

Back to my very earliest memories, my life has been preoccupied with nothing but horses. By the time I was four I was riding a pony to school and I started hunting at the age of five. I used to hunt on a little black pony called Black Beauty who would go anywhere and jump anything. In the early days this meant that I was on the ground as often as I was on Black Beauty, but by the time I was eight, Beauty and I would think nothing of jumping a five-bar gate out hunting. A year or two later, if hounds were near on a Monday or Thursday, I would develop a convenient sick headache at school, before break, so that I could be sent home and have a good day's hunting. This for a season and a half worked wonderfully well – it was simply a question of riding Beauty to school, being taken sick (usually by ramming two fingers down my throat), being sent 'home' and going hunting instead. But after a couple of seasons, I inevitably got found out, had my backside well and truly tanned and after that my hunting was confined to Saturdays.

One day I remember we were hunting at Barrington, and Barrington Park was divided neatly by iron railings, each of them only about three feet six high, but considered more or less un-jumpable because the horses failed to see the top rail. Hounds found a fox and away they went. I was on the corner of the cover where the fox went away, so giving the hounds plenty of room I set off at a good long gallop, which was Beauty's best speed. It wasn't particularly fast, but adequate for keeping up with hounds.

As we approached the first set of railings Beauty looked at them, sized them up and, deciding that this was a most un-pleasant obstacle, dropped back to a trot. I banged her hard with my hunting crop and she trotted to within two strides, cantered

7

the last two strides and just popped over, putting in a hell of a big jump.

She jumped the three foot six, clearing the railings by a good six inches. I, unfortunately, cleared them by a good four or five feet. I went head over heels, landing in front of Beauty, who stopped and looked at me and asked me what on earth I was doing lying on the ground when hounds were running. I scrambled back into the saddle and away we went.

The huntsman – Oliver Moss, who, apart from being an outstanding huntsman, was an excellent horseman – had also jumped the railing but no one else had dared to try, so I put Beauty on the tail of his horse, a famous hunter called West End, and he and I had five or six sets of railings by ourselves, which took us completely away from the field.

This was the first time in my life that I had ever been more or less alone with hounds. The pleasure was such that I have never forgotten it, and whenever I've been hunting since I've done my best to leave the chattering mob behind and enjoy the bliss of riding across country after hounds by myself.

I shall always remember Oliver Moss with great fondness as an extremely kind man who indulged a small boy by allowing me to follow where he went. I can only have been a great nuisance to him, but he put up with it and encouraged me in every way. In 1939 he went back into the air force – he was an ex-airman – as one of the test pilots of the first of the Gloucester Meteors. He was killed on a test flight in 1943 or 1944.

As well as playing truant for the odd day's hunting, I was also riding to school every day. My father was doing a little bit of horse dealing on the side, buying the odd pony which might or might not make a show pony or a show jumper or a gymkhana pony, and I was usually the first to be put on top. My elder brother Charles, though a much better horseman than I, preferred to stick to his own ponies which he was showing and show jumping, rather than try some impossible animal that my father had bought cheap off the gypsies or in Exeter market.

These odds and ends of ponies taught me some things very quickly; that there must be an easier way of teaching a horse to be ridden than hitting the ground with a bump over and over again until the pony got tired of depositing you; there must be an easier way of teaching it to turn left or right than catching hold of one rein with both hands and heaving it round; that there had to be an easier way of catching a horse than chasing it round and round a field until he lay down from fatigue; there was surely an easier way of teaching it to jump than pointing it at an obstacle, saying the General Confession and the Lord's Prayer and hoping for the best. I discovered that if I could get the horse to want to do what I was trying to get him to do, he would do it.

For example, it was much easier to teach a horse to go from a walk to a trot going home than going away from home; and it was easier to teach him to turn to the left or right if he first learned to respond to the rein and the heel by following another horse. I also discovered that the most important thing of all was to make my wishes clear to the horse without annoying, frightening or upsetting him. If I was relaxed, happy and comfortable the pony would be relaxed, happy and comfortable.

This was the beginning of my study, which developed over a period of thirty-odd years, of how horses communicate. Because to get him to do what I wanted him to do, it was necessary to communicate my ideas to him, it was important to know how he thought and what made him want to do certain things and dislike doing others. Also as we went on I could see that some horses liked doing one thing and hated doing another.

It became very quickly obvious, in short, that horses are as different from each other as chalk is from cheese. Whilst they may look something alike, their thoughts, their desires, their abilities vary with their temperament and experience.

At this time in 1939, one of the best show jumping ponies we had was a little thirteen-hand Exmoor mare called Susan, who was kind, sweet and rather sleepy. You could put any fool on her and she would walk and jog along quite happily without putting a foot wrong. If you took her out hunting you had to work like hell to keep anywhere near hounds at all. Yet the same pony, as soon as she got into a show ring, altered completely –

she became alive, alert and jumped superbly. Beauty, on the
other hand, would deposit you on the ground with absolute
regularity once or twice a week, just to tell you to mind your
manners. Out hunting she was a marvellous ride. As I've already
described, she would jump anything, and provided she was in
sight and sound of hounds she was happy. But take her into the
show ring and canter into the first fence, she would prick her ears,
lengthen her stride, and just as you got ready to take off she
would stop. You would end up, to the delight of the crowd, on
one side of the fence while she would stay firmly on the other.
If you let go of the reins she would go back to the collecting ring
and say 'Well, that's it for the day' – and it was.

Beauty didn't mind hunter trials so much, although she was not
very reliable. She liked working in pairs, with my brother's horse
Bill the Baby, because the pair of them used to enjoy racing each
other. So we did quite well with her in the hunter trialing in the
pairs events.

Here you have the perfect example of one pony who was something of a sluggard at home and a superb show jumper, and the
other a superb pony at home or out hunting who wouldn't have
anything to do with show jumping. This was my first experience
of how horses vary in thought, character and what makes
them want to do a certain thing – in other words, motivation.

The more horses I handled, the more I learned to take account
of these widely varying characteristics. And I also learnt that,
whilst you may be able to improve a horse's particular abilities
with training, and you may also change the things he wants to do
by handling him, and teaching him to do something else, in the
correct way, you only have to make a single mistake by teaching
him in the wrong way, to end up with a bloody-minded creature
as stubborn as a mule.

I also found that there are certain horses who want to do
something but are physically incapable of doing so. Old Fearless
was such an example. She was an extremely bad jumper and not
particularly good looking, but we found that she really enjoyed
the showing.

We took her to one very small show and gymkhana, tarted her

up, plaited her mane, groomed her and she got a third in the show class. But she had every fence down in the show jumping and was hopeless at gymkhana work. The following week I noticed that she went about her work with an enthusiasm that she hadn't shown before. Then came the following Saturday, when we were off to another gymkhana, but on the basis of Fearless' failure the previous week we left her behind.

On Sunday morning I took her out to pull the dung up to the field. When I went to harness her up, the first thing she did was take a swipe at me with her hind leg, and when I tried to put the collar on she fastened her teeth round my shoulder and shook me like a rat. Eventually I got her harnessed in the dung putt and started to haul the dung up to the field. She wouldn't pull at all, and she was awkward the whole of the following week.

So, come Saturday and gymkhana day, I groomed her and got her ready again, and rode her leading a string of other horses to the gymkhana. I left her tied to the fence, just putting her in for the show class, where again she got a second or third. Then we rode home again. Her behaviour that week was transformed. She did everything with enthusiasm and, apart from the odd nip when I annoyed her, her behaviour was angelic. I very quickly came to realize that she looked upon her Saturday off, going to a show and mincing around in the show class like a pimp in Piccadilly, as a reward for a hard week's work. If she didn't get her reward, God help anyone who tried to handle her.

She wasn't a very good hunter, either, but I had to hunt her occasionally to keep life and limb intact. Shortly after this I smashed my leg playing rugger, and, as soon as I was capable of doing so, I went hunting in a dog cart with Fearless between the shafts. This was the thing which really suited her because I wasn't trying to change the habits of a lifetime; so Fearless and the dog cart were near the front of the field, and when hounds went we went, hell for leather across country, from gate to gate and gap to gap, with the cart bouncing up and down like a jack-in-the-box.

Although I smashed my plaster regularly, Fearless and I really enjoyed ourselves. It was not, however, a pleasure necessarily appreciated by the odd-bod who asked for a lift. I remember one occasion when hounds went away while I was giving a friend a lift in the trap. We galloped up the headland and across a ploughed field, with the cart bouncing from side to side, through a gateway, only to find that hounds had gone away through a small hedge into the next field. I had the alternatives of going back or jumping the hedge, cart and all. So I caught Fearless one across the backside with the loose end of the reins, and we went into the low bank, which was about a foot high with a two-foot fence on top. We hit the low bank at a rate of knots, bouncing up into the air and landing on the other side – fortunately with the two wheels still on the cart and the two people still inside it. We cantered down the field and got to the road, where there was a crowd in the gateway. My passenger scrambled out without even having the good manners to say 'thank you' for the lift!

After about two months of this there wasn't much left of the cart, and my leg hadn't got much better. But Fearless's work rate had gone up fifty per cent. The intelligence of the mare was such that if she walked into the yard before lunch and saw me washing the mud from the previous week's hunting off the trap, she knew that we would be going hunting on the following day. And after lunch the person who was working Fearless was going at a rattling rate, with no hope of stopping her.

She was also one of the many horses I've had who actually enjoyed being tarted up and made pretty. Normally when you wanted to groom her and brush the mud off, she'd be trying to catch you a sly one with her heels when you were at the back end, and having lumps of backside for breakfast if you were near the front. But when you were preparing her to go hunting or to a show she'd stand like a rock and you could groom her anywhere and do anything with her and she'd never move. When you plaited her mane she'd put her head down to the ground so you could plait in the easiest position.

Another of the things I noticed was the way that some of the horses enjoyed doing certain tasks. For example, Champion the cart horse adored horse-hoeing. This entailed pulling the horse-hoe, which was an implement something like a plough with two blades set at an angle sixteen inches apart, between two rows of mangels or kale planted eighteen inches apart. If the horse deviated, instead of hoeing out weeds you hoed out a neat line of kale leaving a horrible gaping hole. Usually when doing this you needed somebody to lead the horse and another person to guide the hoe. Not so with Champion. The precision of the work so fascinated him that he would lift each foot, placing it precisely in position, so that the person guiding the hoe had neither to steer the hoe nor Champion. Champion's pleasure in doing a difficult job precisely was a pleasure to see, and his motivation to complete the work perfectly was so great that if you attempted to steer him out of line he would ignore the reins completely.

In direct contrast was Caravan. On the one occasion we tried him I led him for the first two rows and he danced about all over the place, leaving more gaps than kale. Then my father decided that he would lead him. Caravan was very angry by this time, and after doing half a row – during which he flattened my father's feet with his own – the old man called over my sister, Olive, who had come to watch the fun. She was set to lead on one side and I on the other. We got ready to start, my father took a firm grip of the reins and the horse-hoe and told Caravan to hold fast, which he did by plunging forward, throwing Olive and me aside like wisps of hay, and at the same time jerking the horse-hoe and reins out of my father's hands. He proceeded around the field at a good long gallop, dragging the horse-hoe across the ground in a series of kangaroo leaps. When eventually he stopped my father gave up an obviously impossible task and went to fetch another horse, leaving me in punishment to fill in the gaps left by Caravan with spare kale plants, a much duller and more back-breaking task.

My father, commonly known as 'the Boss' or 'the Old Man', had a very great influence on my handling of horses and on my view of life in general. A born rebel, he stood six feet and was built like a boxer. He had a temper which exploded like dynamite and which had the instability of nitro-glycerine. He had black, curly hair which didn't change colour until he was over sixty. He never believed any statement or theory until he had proved it for himself, and although a deeply religious man he questioned everything and thought it out for himself. As far as the law was concerned, if he didn't agree with it or thought it stupid he would disregard it completely. One of his creeds in life was that you should never ask anybody to do something you wouldn't do yourself. Whilst at times hazardous, life was never boring with him around.

His love for animals was very deep and to him horses were, to quote his own words, 'God's supreme creation'. Whilst if he thought it deserved it he would give a horse a hiding (though not quite as severe as the ones he gave me), he could not stand to see an animal ill-treated. I once saw him knock down two gypsies who were ill-treating a pony, take the pony out of the cart, put me up on its back and take it home, the two gypsies following on behind and finally getting more out of my father for it than the pony was worth. One of my final memories of him was seeing him going into the kitchen, aged about sixty-eight, covered in mud and with a grin splitting his face in half. 'I thought I was getting old', he said, when I asked what had happened, 'but I've just had the hell of a fall and haven't hurt myself a bit, so I can't be as old as all that.'

He vividly demonstrated those qualities that are most import-ant in handling horses – determination, humour and endless patience.

The urge to do a job perfectly is extremely great in some horses. Others just want to get the job done as quickly as possible not minding how the hell it is done.

Another careful worker was Bonnie, although she walked with a limp. As a yearling she slipped a stifle and my father let her run on for another two years for the stifle to heal, but as soon as he broke her in to ride she bucked and the stifle came out again. She was left for three or four months, but it didn't heal and we were left with the unpleasant choice between using her limping, going lame on one hind leg, and having her put down. There was no choice, of course. She was happy and contented and she wasn't in pain so we worked her in harness, walking. The one job she loved was hay-making. The horse had to learn to stop at a precise spot, so that she left a neat row of hay behind her. As the driver was busy operating various levers, it didn't leave much time to drive the horse as well. So once you had gone up the field and dropped the first bundle of hay in each line, Bonnie would check at the exact spot that you had to drop the hay coming back, and so on right down the field. She would do this all day with her ears pricked, watching you and watching her lines and checking at the exact spot the whole time. The pleasure she got from doing the task perfectly was very plain to anyone. Of course, she was able to do only light work about the farm, but she did that well and enthusiastically, especially neat and fiddly little jobs that the other horses hated.

This gave me further proof that each little job the horse was doing could be a reward in itself, and that a large part of the successful handling of horses lay in finding the task that the horse did well and enjoyed doing.

2: *The Mind of a Horse*

My observations about the importance of understanding each individual horse's personality and behaviour, what one horse liked and another did not, came home to me even more forcibly when I started racing. It was soon apparent to me that horses of lesser ability were beating horses on the racecourse who were much faster and better jumpers, simply because they wanted to race and were determined to win, whereas the horses they were beating didn't want to race, so as soon as you got them up front they'd almost stop. You could see that some horses, as soon as the starter's flag was dropped, would take a hold of the bit, tear the reins out of your hand and go as fast as they could for as long as they could; whilst others, like Old Doleful, would lollop around half asleep for the first circuit, and only after about two miles get warmed up enough to start galloping. So the variation in the psychological make up of the horse is one of the important factors in racing.

What became equally obvious in observing the horses was that some horses would do anything for you, while dealing with others was like trying to get a response out of a block of wood. This in its turn led to the conclusion that there's more to communication between horse and rider than pulling a pair of reins and catching him a clout with a stick.

The third thing I found out was that teaching a horse to do something was a great deal more complex than ever the 'experts' lead you to believe. My reading told me that learning in a horse was merely the result of repetition of the same tasks over and over again. But my observation showed me that if you tried to teach a horse to do something in one way, he would learn much more quickly than if you taught him to do the same task another way. I needed, in fact, to know a great deal more about the process whereby learning in horses takes place.

16

The horse that first set me thinking in this direction was the same little pony that took me to school, Black Beauty. Now she could get any halter off, untie any rope, open any stable door and open most of the gates on the farm by using her head, her neck and her teeth. She could undo knots, pull back bolts or lift gates. The only gates she couldn't open were the gates that were falling to pieces and were tied into place with a dozen different pieces of string, and this was only because they were too heavy and awkward for her to shift. They were in any case no problem to her because she just jumped them. But it struck me very forcibly that she had learned to do these things simply and solely by teaching herself. And she had taught herself to do these things because she wanted to wander around the farm where and whenever she pleased. This desire was so strong that she had learnt tasks that the other horses found impossible. The corollary of this was quite simple – teaching a horse to do something new would be much easier if you could make the horse *want* to do it.

This conclusion was reinforced by my observation of polo ponies. A pony who enjoyed the game, I found, would require fewer and fewer signals from me, until it was merely a case of shifting my weight and touching his mouth to get him to do what I wanted. And even if the pony got a clout from a polo stick, which happened from time to time – not only did the pony get clouted, I got clouted by a polo stick once every week or so – it didn't diminish his enthusiasm for the game.

But to get a horse keen on something I had to teach him first of all what it was I wanted him to do: or rather, what I wanted him to want to do. And this involved learning something about horse communication. If I could learn how horses talked to each other, I would be in a better position to convey my own wishes to them.

Once I had discovered that the common idea that animals communicated using particular sounds to express particular feelings or concepts was erroneous, and began to watch how my horses conveyed their wishes and desires to each other, learning how they communicated was only a matter of study and time. I discovered that it was the tone of the vocal sound that mattered, not the sound itself; and that sign language was very important. But as well as this, I found that horses can convey their feelings and desires from one to another over great distances without using either sign or sound – just as if I was excited my horse would be excited, and if I was cool and calm I could make my horse relax. This form of communication I identified as a form of ESP, which we eventually extended to include communication by telepathy – conveying mental pictures, as well as feelings to our horses.

The results we obtained from using ESP as we developed our abilities were really quite amazing. The original work was done during training, but we found it equally useful in competition and general riding.

One of the most dramatic examples of the success of ESP was with a little Welsh cob I had. In those days transport was extremely difficult to get hold of, petrol being short, and when I was going to a hunter trial, show or gymkhana, I used to ride one horse and lead the ones who were competing. They arrived at the competition as fresh as was possible.

I rode Witch to the hunter trials at Cattistock, which was about fifteen miles away, leading three other horses with me. On the way there she behaved like an absolute cow, dancing and jogging all over the place and pulling the other horses from side to side of the road. I was really annoyed when I got there, so I gave the other three to someone to hold. She was still dancing and pulling herself, so I thought I'd get her steady before I got home by giving her a fast mile around the field. I got her settled down and galloping on, but as I came to the fence she'd no intention whatsoever of turning, and we went over the bank at a rate of knots. How she jumped it I don't know, she'd never jumped a fence in her life before, but she found about eight different legs as she landed and went away up the next field. The only way I could get back to the other horses was back over the bank, so I swung her round and put her back at the bank and she jumped it much more efficiently. I decided that since she was there and still much too lively to do anything or give me a comfortable ride home, the bitch could go in the novice class of the hunter trial.

I rode my other novice first and then came Witch's turn. I was extremely apprehensive and somewhat excited and she was as excited as she could be, so we danced down to the start and away we went. I had no hope of steadying her going into the first fence, which was a post and rail, and we went into it much too fast. With me saying my prayers and hoping for the best, we approached it. Ten yards from the fence she suddenly saw the object in front of her and half tried to refuse, but she was going too fast to stop on the slippery ground. She arched over it – this was the first proper fence she had jumped in her life – and away we went into the second fence. This was a straightforward bank, and after jumping two, she was absolutely certain that she was an expert on banks and I was reasonably confident. Over we went, getting between the flags by the skin of our teeth, since she was veering to one side. We had to turn at right angles to the next fence and to my surprise, instead of going twice round the field before I got her facing it, she swung round like an old hand and, having got the worst of the steam out of herself, she went into the gate at a good steady gallop. And so we proceeded round the course to the twelfth fence which was an extremely difficult one – you had to pop over a post and rails, go down into a quarry and jump out over another bigger post and rails at the bottom. Everyone had been having trouble with this, I'd already had one refusal on the other horse and only about two people had done it correctly and got through without refusing. I galloped Witch into it – or rather she galloped herself into it fairly fast – and as she got close she could see that she was more or less launching herself straight out into mid-air and she steadied back to a trot, popped over the first one without any difficulty, landing a little far out so that she slipped down the quarry half on her bottom. As soon as she reached the level bit at the bottom, since her hind quarters were under her, she gave a terrific spring and went over the second post and rails like a rocket. We finished the course to find that we were in fact one of the only three clear rounds of the day and I was in the jump off. But having completed the course successfully once, I decided that the little mare had done enough. By this time she had changed from being, 'that bloody-minded little bitch' that I had to hack to the hunter trials, into a good little mare and we were the best of friends.

Riding home in the half light of the evening, over fifteen miles, I had time to think about Witch's performance and analyse it. Three things were very plain, (a) she obviously had terrific natural ability, (b) she had changed within an hour or so from something that was only half broken and almost uncontrollable into an easily controlled horse and (c) her desire to gallop and jump, correctly harnessed, had turned what promised to be an extremely difficult ride into a very good ride indeed.

The thing that surprised me most of all was that because I knew my reins were more or less useless as she was more or less impossible to stop and steer, I hadn't bothered to use them, and simply by using the weight of my body, and by looking from fence to fence – i.e. using ESP – I had managed to steer her round the course. She had responded to the slight signals that I had given, but mainly to the oneness that had grown between us. But this had been possible only because she had clearly been enjoying herself, and could see the object of what she was doing. This led me to the conclusion that if you could discover the right motive for getting a horse to do something, the horse would do it with very little difficulty at all.

So, not only was I working on how best to communicate with my horses, I started studying how their minds worked – where I was successful in doing this, training became much easier, because it was merely a question of discovering the physical and mental needs and desires of the horse, and then harnessing them to the ends I wanted.

But the third leg of the triangle also had to be worked on. And that was discovering how a horse learned, other than by dull repetition. My experiences made it obvious to me that dull repetition was extremely tedious and very hard work for me as well as the horses. If I could harness my understanding of communication and motivation with some comprehension of the process whereby horses learn, I could not only make it easier to break and train a horse, but when it came to competition I could increase my horse's ability, power and will to win; or at least to co-operate in what I always wanted to do, which is to get a horse performing above his normal ability at the correct time and in the correct way.

3: *How Horses Learn*

The thing that sets the horse apart from most other domestic animals is his incredible capacity to learn. But it is the extent and kind of his learning that makes each horse into an individual being. What he's learned and how he's learned it reveals itself in all his behaviour. Through his learning capacity, combined with his ability to communicate and be communicated with – which is itself extended by learning – he acquires his habits and customs and develops his abilities. Since everything he does derives from what he has learnt, this is the key to understanding how the individual horse behaves.

The young horse may learn by imitating his mother; for example, he learns what to eat partly by trial and error, and partly by imitating and selecting food his mother eats. Early in his life he develops an attitude to man which is directly connected to the way man has behaved towards him, and his mother before him. Thus if in his early contact with man he has been ill-treated, he will after that instinctively distrust man. And if his mother has a fear and distrust of man, this will be communicated to him. I came across an example of the power of early learning while I was working on this book. I had a desperate message from a friend who lives just outside Lampeter. Would I come and help him catch his mare and foal, because he wanted to take them to Llanybyther horse sale. When I got there I discovered that the mare, though extremely nervous, was not too difficult, but the foal was absolutely terrified of all human beings. This was so unusual that I asked the owner why. It seemed that kids from a nearby housing estate had been throwing stones at the horses to make them gallop. This had gone on for the past three or four months, and had made the foal absolutely terrified of all human contact. By the time he had discovered what was happening the foal was so wild that he couldn't move it to a quieter part of the farm since he was unable to get near either the mare or the foal – they galloped round and round the field in circles rather than go out of the gate. Finally he decided to solve his problem by selling them.

We were very fortunate that there was only a short, quiet lane from the field to the farm. So we solved the problem comparatively easily. We took the owner's small daughter's pony and left it with a bucket of nuts just outside the gateway and then walked quietly down to the far end of the field, which meant that the foal tore up the field and the mare followed at a quieter pace. When they got to the gateway they saw the pony eating out in the lane. They stood watching the old pony eating for a while, then the mare walked slowly towards the gate, stood for a minute, walked through, followed by the foal, and drove the old pony away from the bucket of nuts. From there we drove them slowly into the cattle yard.

It took me about two hours before I eventually managed to get a hand on that foal. It literally shook with terror, but slowly and gently I got it quietened down just a shade, and after another hour and a half it was beginning to trust me a little bit. I stroked it gently, imitating the mother nuzzling it. And since it was a lovely cob filly it seemed a pity to send it to Llanybyther and the owner didn't take much persuading to keep it. So we loaded the mare and foal up into a trailer and took them to a quiet field at the far end of the farm, hoping that by degrees, by the time the foal was two or three years old, it would have quietened down and lost its fear of human beings.

If, on the other hand, in his early contact with man the young horse is allowed to misbehave and walk over the top of whoever is handling him, he will never develop any respect for mankind whatsoever. Over the last ten or fifteen years we have had thirty or forty horses sent to us for gentling, as unbreakable. And by far the greatest number of difficult and unmanageable horses we have seen have had this type of upbringing. Horses that have been treated like mischievous puppies when foals, and allowed to get away with biting, kicking and pushing their owners around because at that stage they're such playful, pretty little things, by the time they're yearlings, are thoroughly spoiled and badly behaved. When we get them as three- or four-year-olds, they have had no discipline whatsoever and are used to doing exactly what they want. If the owner does not move aside when the horse pushes past him, the horse bites or kicks until he does, and ends up with no respect for the human being whatever. Such horses treat their owners as subservient members of their own herd. If they tell them to get out of the way and they don't, they immediately kick or bite them. This is normal equine behaviour. This behaviour has all been learned. And if the horse is to have any kind of future, it has to be unlearned, and better behaviour substituted. How can this best be achieved?

How a horse learns is a very large subject indeed. But the study of horse psychology which is in effect the study of learning in horses, and how this knowledge may be applied in handling, can help us to be clear about what we are doing. To begin with, what is learning? It is a relatively permanent change of behaviour which occurs as a result of experience, practice or training.

This definition has three important elements. It says first that in learning there is a change in behaviour. This may be for better or worse, but if there is no change, there is no learning. The change in behaviour may also not show up immediately: if for example you are trying to teach your horse to stop on a word of command, the first time you say 'whoa' he will take no notice of you whatsoever – you'll have to stop him by other means. But, by degrees, if you quietly make him stand still every time you say 'whoa', after eight, a dozen or maybe twenty times the horse will stop automatically. Then we say that the horse has learned to stop on command. You can see the difference – his behaviour pattern has been changed by learning.

Learning can be any form of change of behaviour as the result of experience. It need not be the result of conscious training. In teaching a horse to stop on command you are changing his behaviour pattern by consciously training him to do something. But the training could be completely unconscious, the result of an outside influence – such as the weather. For example, a horse will automatically turn its backside towards wind and rain. This too is the result of training; but no one has rushed out every time it started to rain and turned the horse so that its bottom is facing into the wind. It is merely that over the course of his early life he has discovered that if he faced into the wind and rain, the wind and rain blew into his nose and eyes, so he turned away from it and pointed the part of his body that is least affected by wind and rain towards it. This is a form of learning and a form of training: it is unconscious learning, and the training has been done by Mother Nature, but nevertheless it is as much learning as when a horse learns to jump a four-foot-six fence.

The second part of this definition is that learning takes place *through experience or practice*. Other changes of behaviour, such as those which occur as he gets older, or through fatigue or injury, do not count as learning. If a horse stops because he has injured himself or because he is too tired to go on, this is not learned behaviour.

Third, the change must be relatively permanent. If it is not, it is probably due either to a transitory change in motivation, such as fatigue, or to adaptation. For example, if you move a horse abruptly from light to darkness he will immediately lack co-ordination and be unable to see, but in a very short time he will adapt to the darkened conditions and be able to move around with comparative ease and freedom. This is not because he has learned anything, but because his body has adapted to the darkened area. Here it must be stressed that such stimuli as fatigue, restriction of exercise or overfeeding may *also* lead to learned behaviour. For example, a horse that is repeatedly overworked or overtired will learn to become lazy, a horse that is kept underexercised may learn to become excitable or shy, or a horse that is overfed may become completely unmanageable. In cases such as these the horse will have to undergo a long and laborious course of retraining. This is particularly difficult, because first the horse will have to learn to forget its previous experience, and then will have to be retrained to regain its previous level of performance.

According to behaviourist theory, all learning is the result either of classical conditioning or of operant conditioning. In both a specific response to stimulus or stimulus conditioning is required. Classical and operant conditioning differ first by the nature of the stimulus, second by the kind of response learned and third by the nature of the response to reinforcement.

Conditioning means quite simply placing an animal or a human being in a situation likely to make him respond in a certain way : to subject him to a certain stimulus. And stimulus may stimulate an animal in either of two ways : it can stimulate him positively, i.e. make him want to do something – make a horse want to please you, for instance, or make him want to get home for dinner; or it can stimulate him negatively, i.e. make him want to avoid something – he moves away from a slap on the backside, or avoids incurring your anger. To both these types of stimulus he will respond by doing what you wanted him to in the first place, but the learning process can be very different.

For example, suppose you want to catch him, you call him and he comes over to your handful of nuts or oats. He learns to respond to your calling because he knows he will please you by coming to you and he'll also get a handful of nuts for doing so. So, the stimulation and response are directly connected with each other : the call stimulates him, the response is to come to you and he'll immediately be rewarded by being caressed, made a fuss of and given the nuts. We say then that he is conditioned to come to your call. The nuts and the patting are scientifically known as the reinforcement, since they reinforce the learning. If you don't pat him and give him the nuts he is less likely to come over next time you call him in the field, and in a very short time he won't bother even to look up from his grazing when you give a shout. This is a simple example of conditioning, or training : and as in all learning, it involves stimulus, response and reinforcement.

It is easy enough to illustrate the difference between classical learning and operant learning, with another example. When I'm feeding the horses I measure out the feed into the basins in the feeding house, dropping the first one, which is usually

Cuddles', first and then the second one and the third one and so on, outside the stable doors, putting the feeds in as I go back up the line. It's a very simple, quick and efficient process, but Cuddles gets his food last. This of course doesn't please Cuddles very much because (a) he's extremely greedy and (b) he's convinced that he's the most important horse on the place and therefore should be fed first. For a long time he tried to reach over the top of the stable door to get at the feed in the bucket.

On one particular morning, however, either he wasn't particularly hungry or he'd got a very bad itch under his chin, because instead of reaching out for his feed he scratched his chin on the most convenient object which happened to be the bolt of the stable door. In so doing he pushed the bolt back, the door swung open, he walked out and started his feed. Within three days he was pushing the bolt back with his chin and opening the stable door whenever I was feeding. Another day or two and he was opening it whenever he felt like it. So we pushed the catch down so that he couldn't open the stable door. When he found that he couldn't open the bolt by rubbing it with his chin he fiddled about with his lips, released the catch, opened the door and away he went again. The next stage was to put a padlock on the door. But since by nature I'm careless and lose things, I used to leave the key in the padlock, and he learned not only to turn the key in the lock but to remove the padlock from the latch and then open the latch so that he could let himself out.

This is a typical example of operant learning. At first by chance he did the correct thing, i.e. pushed the bolt back, and got an immediate reward with food. Then from learning to do it for food, he learned to do progressively more difficult tasks, each for a reward. In the beginning the reward was early breakfast, but later on it was freedom to wander round, come down and see what we were doing in front of the house, tease the other horses and make a general nuisance of himself. And of course the greatest reward of all was that he could annoy me, and when he saw me getting particularly angry he would belt back into his stable and pretend he'd been there all the time and was an innocent little horse who'd never misbehaved in his life. His behaviour had changed from that of a horse that could quite safely be shut into a stable by closing a bolt, to that of a horse who could open a stable door and go in and out whenever he wished to. The difference in behaviour pattern is quite easy to see.

The way a horse's behaviour pattern can be changed temporarily, but not learned, is equally easily illustrated, again using something that happened to Cuddles. We were out hunting near Lampeter and had to go down a steep and slippery hill. He slipped on his hind legs coming over a fence, and his hind legs shot forward underneath him so that he slipped the whole way down a hundred-and-fifty-yard slope, sitting on his bottom. It is of course completely contrary to any horse's normal behaviour, to move a hundred and fifty yards sliding on his bottom like a small boy on a toboggan. When we got to the bottom of the hill, however, he got back to his feet as if nothing had happened, and away we went again to have an extremely good hunt.

Normally Cuddles, when he's with another horse, tends to dance along with his head in the air, jingling his bit and looking at everything and generally enjoying himself, but that day, because we'd had such a hard hunt, he was extremely tired and we hacked the two and a half to three miles home with Cuddles plodding along like an old nag, with his head hanging down half way between his legs. And this, like the toboggan slide, proved to be only a temporary change of behaviour, in this case due entirely to fatigue. Since then he has never moved across country sitting on his bottom, nor, except on the odd occasion when he has been extremely tired, has he plodded along like a tired old nag. Both of these temporary changes in behaviour were due to *unlearned* behaviour patterns.

We have given an example of operant conditioning, in Cuddles' discovery of how to open his stable door. Let us now look at classical conditioning. In classical conditioning, first the stimulus is a specific event – such as a flash of light, tone or note which is briefly present, your voice giving a command or the touch of your heels. Whereas in operant conditioning the stimulus is not a specific event, it is a longer lasting situation which has several features, only one or two of which prove relevant for learning.

Second, in classical conditioning the response, like the stimulus, is a specific one. Moreover, it is usually a reflex or an innate reaction. In operant conditioning on the other hand the responses are at first varied and random.

Third, response to reinforcement differs between the two forms of learning. In classical conditioning the reinforcement – the reward or punishment – is always part of the conditioning situation, regardless of what the person or animal does. It does not depend upon the response made.

A simple example of this kind of reinforcement is used in teaching a horse to stop when you say 'whoa'. In training a horse to stop when you say 'whoa', you pull on the bit with the reins so that the horse stops, which relieves the pressure on his mouth. So, the stimulus, putting pressure on the horse's mouth with the bit, is immediately followed by the horse stopping, and as soon as he stops the stimulus is removed, that is, he is rewarded. This is a form of classical learning.

But in operant conditioning reinforcement does depend on the response. If the subject does the right thing he is reinforced positively, that is he is rewarded with titbits or by other means, if he does the wrong thing he is reinforced negatively with a punishment.

A simple example of the operant process can be used to deal with a horse which refuses to lead. The person leading the horse has a basin of food from which, if the horse leads forward, he receives a reward. If he refuses to lead forward, an assistant throws a series of well-directed missiles at his bottom until he does move forward. Here you are teaching the horse to respond to positive reinforcement, that is the food, and to avoid negative reinforcement, that is the missiles. So, in this way he is conditioned to lead. I would add that it is infinitely better to throw small pebbles at his bottom than to catch him one with a hunting crop, since you're much less likely to have a mouthful of horse's hooves as an unwanted meal.

It will be seen that a feature of operant conditioning is that the response to the horse's behaviour should be immediate, whether the response is positive, negative or both. If a horse jumps a fence, you make a fuss of him. If he doesn't jump the fence you smack his bottom, so you can see that the reinforcement depends on what he does and if he does it.

I have had to put a lot of emphasis on the two different kinds of learning, because when you are training a horse it is extremely important to know which kind of learning you are going to use. The drawback to classical learning is that if the stimulus is insufficient it will be ignored completely, so the stimulus may have to be strengthened from time to time.

Suppose you are riding a horse. If you drive your heels into the horse hard he will shoot forward. If instead you just tap him with your heels he will move a little bit faster to begin with, but you will end up having to kick, kick, kick while the horse progressively ignores you. So though classical training will improve performance slightly, it can lead to your over using it, and in a competition for instance you may end up in front of the judges and stewards – and rightly so – for ill-treating your horse. The point is that in training a horse by classical conditioning the emphasis tends to be on punishment, and this is something I abhor very strongly. Punishment, of course, is necessary from time to time but over-use of punishment is never a good thing. In fact, punishment over a long period leads to a deterioration in performance.

To illustrate this, I knew a very promising young show jumper, who tended just to tap a pole. To cure this, her owner covered the top pole of several of the jumps with hedghog skins, so that if she hit the top pole she would prick herself – a very common trick among show jumping people. But Blodwyn, being a very intelligent mare, very quickly learned that the poles in the show jumping arena itself didn't have hedgehog skins nailed to them, so while she would jump everything perfectly at home, taking no chances whatsoever, as soon as she got into competition she would have one or two down.

Negative reinforcement in classical conditioning always tends to lead to this kind of evasion, that is to say the horse will either do just enough to avoid punishment or else will find a way of avoiding punishment altogether.

Bay Star was another example of the counter-productiveness of classical conditioning. She was a very excitable mare when she came to us, with a tendency to rear straight up on to her hind legs and go over backwards. She also arrived equipped with a bridle which contained what might be described as half an ironmongery shop in place of a bit. I quickly changed this to a rubber snaffle bit and took her out exercising. She didn't go up on her hind legs once, although she carted me all over the mountain a couple of times, this being somewhat hair-raising as the mountain is a mass of humps and hollows and bogs and God knows what else! On the first occasion we were motoring along at a fair pace with me vainly trying to stop her, when she went into an old mountain dip, which meant she went head over heels. This didn't worry me very much, but it did worry her and she went very gently for the next four or five days. Then she got away from me again, this time not quite so wildly but it had been raining heavily on dry ground for the previous twenty-four hours. I let her go in a straight line for a while and then tried to pull her to the left. She set her jaw, not having any of it, so I heaved with all my might but she set her jaw even more firmly. I went on like this for about twenty yards and then I suddenly pulled her head round the other way. The pressure that she was exerting to the left was immediately reinforced by my bodyweight, which swung her round sharply. Her forefeet went from under her and again we hit the ground.

Within a fortnight she was cured completely, both of carting me and of going up on her hind legs. The reason for her rearing, it appeared, was that she hated standing still. Her previous owner had been making her stand still while she was out hunting so that he could stop and chat to people. Simply by keeping her moving, I stopped her rearing and this sweet little mare quickly learned that I wanted to go as fast as possible when we were going; but that when I said 'steady' or 'whoa' there was a definite reason for steadying.

So by working in this way, by removing her desire to rear, we cured her completely. We substituted for the classical conditioning – when she moved her mouth was hurt – to which she had been subjected, the use of the mare's natural inclinations to train her. So her training changed from being painful and unpleasant to something that she could enjoy. In actual fact, within a month I had taught her to stand still when I asked her to, and she would stand still quite happily. When she started fidgeting I walked her round a bit and she would be quite happy to stand still again. Here was a mare who, psychologically, needed movement to relieve the build up of tension within herself. When that movement was restrained she reacted by going up on her hind legs.

4: *Using Conditioning Theory: 1*

Classical conditioning gets its name because it was the first kind of conditioning to be studied experimentally, by the pioneer Russian physiologist Ivan Pavlov.

In studying the role of saliva in the digestion of food, Pavlov found, as scientists often do, that something was getting in the way. Salivation was occurring before the food was placed in the mouths of the dogs that he was using as experimental animals. The normal reflex response is that salivation takes place after an animal has taken food into its mouth. But Pavlov found that simply by bringing a dog repeatedly to a standard experimental situation, he could cause it to salivate. Realizing that some kind of learning was going on, Pavlov decided to make a systematic study of it.

For the conditioning experiment he placed a dog in a sound proof room with a one-way screen, so that he could see the dog but the dog would not be distracted by watching him. He hooked up a pan that would be swung in and out of the dog's reach, and he connected a bell to ring inside the room. Pavlov would ring the bell and then, after a few seconds, present the food. He continued pairing the bell and the food together, while measuring the amount of saliva the dog secreted in response to the bell. In this way Pavlov could chart the course of conditioning, or training, in the dog.

He discovered that there was a definite pattern connecting the amount of saliva secreted, and the length of time the dog had been in training. It finally reached the point when the dog was secreting as much saliva at the sound of the bell alone, as would be required for the digestion and eating of its food. The mean curve that was produced was then considered to be standard for all dogs, though of course some dogs had a much sharper curve because they learned more quickly and others had a much flatter curve because they were slow learners.

It is quite possible similarly to plot the response in the horse to any stimulus. We discovered, for example, in our communication work with horses, that a food response (the horse saying 'where is my bloody breakfast', by whatever sound or sign it normally used to demand food) can be produced whether or not a horse is hungry. One of the experiments we have done to prove this is extremely easy to reproduce.

If, for example, you feed your horse at one o'clock sharp every day, within a very short time you will find that, at one o'clock, your horse will ask for food. Then, once you have got this conditioned response to your appearance at one o'clock established, if one day you feed him more than he needs to satisfy his hunger at, let us say, half past eleven, you will find, even if he is not hungry because you have fed him already, that he will still demand food at one o'clock. When you give it to him he may not eat it, but he still shows what might be called a conditioned food response at the time at which he has been trained to do so.

Another kind of conditioning, fear conditioning, plays a large part in training a horse. Most horses have a good many fear conditioning experiences and these fears become part of their adjustment – or very often maladjustment – to stable routine and to being ridden.

A typical case of maladjustment through fear conditioning was a three-year-old I worked with. At the age of three he had been sent away to be broken by a woman who was afraid of him, and who ill-treated him severely. When the horse returned to his owner, whom he had previously adored, he was afraid of her, and in fact of any other woman who tried to ride him, though he went quietly when ridden or handled by a man. The fear of the woman who had trained, or rather broken and mishandled him, had been transferred to all women.

This transfer of his fear of one woman to all women demonstrates the process of stimulus generalization. This means quite simply that if a horse is frightened or excited in one set of circumstances or in a particular situation, it will react in a similar way in similar circumstances or situations in the future. If, for example, a horse enjoys jumping just a small fence and then larger fences at home, then comes across strange fences somewhere else, he will expect to enjoy jumping them and will jump them freely. But if on the other hand he is frightened by a car, he will tend to be frightened by all motor vehicles until the fear has been cured.

All that is required, then, to condition fear is to equate some unnatural stimulus with some natural or unconditioned stimulus for fear. An important feature of fear conditioning is that it should happen fast. In teaching a horse in its early stages it must be remembered that the memory of fear lasts a long time, since the learning is much stronger than that which results from positive stimulation. So the use of fear conditioning may speed training considerably.

For example, salivary conditioning in dogs will take a considerable time, but from fear a dog can be taught very quickly. It will take you a week, ten days or a fortnight to train a dog to salivate when you sound a bell. If on the other hand, when you sound the bell you also administer an electric shock to the dog's feet to make it jump a low barrier, at the third or fourth sounding of the bell the dog will jump the barrier before he gets the electric shock.* This effectiveness of fear based conditioning is probably one reason why the term 'breaking' is used for the training of horses – it is easier to train a horse with fear than it is with patience and understanding.

You can see this distinction in various ways of teaching a horse to stop. The way we do it is to ride the horse we are training down the road on the heels of an already trained horse. When I say 'whoa' the trained horse in front will stop, which means that the horse I am riding will also have to stop, or walk into the backside of the other horse. I may have to repeat this twelve times before the horse I am riding will connect the word 'whoa' with stopping. But if, on the other hand, I were to adopt a different technique, that of hitting the horse across the nose every time he didn't stop, he would learn

* Peden, *Fear response in dogs*, 1954.

in one or two lessons. Fear training will always have a much quicker effect than training by example.

Since fear training is quick and easy, why do we take so much trouble to avoid it? The answer is that it is effective only when teaching the animal to perform relatively simple tasks, such as stopping on command. It is a short-term technique.

When you are training a horse you are working with a long-term end in view. You want a finished product, a co-operative, willing and enthusiastic horse, and if you use the method of avoiding pain the animal will very quickly learn the minimum that is required to do so. So you will end up at the best with a horse which is doing the absolute minimum to avoid punishment. At worst, if you are dealing with an excitable and nervous or a very strong charactered horse, you will end up with an animal that is completely unmanageable.

On the other hand, training solely for reward, if not combined with strength of purpose and discipline, will produce a disobedient horse which does only what it wants to. As in all things, a mean between too much severity and too much kindness is necessary.

From our own and countless other people's experiences, if you can get a horse to want to do what you want him to do, you will undoubtedly get the best out of your horse. You want an iron fist, but in a velvet glove. We always go as far as we can saying 'Please will you', and try to get the horse to enjoy what he is doing. But if the horse says 'No I won't' then we make him do what he is told, just using patience, and in the last resort a good hiding.

The perfect illustration of this is Charlie. We bought him at Ascot; he had been broken as a yearling and put into training. He had run as a two-year-old, three-year-old, four-year-old and five-year-old, so for the previous five years had known nothing but the training stable and training routine. From the way he went when I rode him first, I don't think he had ever been ridden except with a string of other horses. When I took him out the day after he arrived, he literally shook with fear at the idea of going out by himself. Without the feeling of security he had from working with a number of horses he was terrified at every gateway, and it took me twenty minutes of patience and coaxing to get him past a forty-gallon drum which had been left by the side of the road.

The following day when I took him out he was beginning to enjoy the novelty of so many fresh things to see, and he took no notice of the forty-gallon drum. But the direction we took that day entailed leaving the road and going down a very slight bank, through a small gateway into the forest. In Charlie's view, my polite request that he should do so was completely outrageous. When I became more insistent he proceeded to buck and kick every time I tried to make him go forward with my heels. After forty minutes, both of us were in a muck sweat and stronger methods became necessary. I got off and opened the small gate on foot, instead of doing it from the horse's back as I did normally. I got back on and turned off the road and down the bank and eventually backed him through the gateway and up on the road the other side.

Immediately we were on the road, I told him how clever and how brave he was; I got off and made a terrible fuss of him, then I got back on again and told him we had to go back through the gate. This was more than he could bear, but after a battle of only about ten minutes, we were through. Then I turned him round and went back through the gate again without very much objection on his part; and the fourth and fifth time he walked willingly backwards and forwards through it.

Normally of course, I would avoid a battle so early in my relationship with a horse. But to be quite honest, since I had been going through that gateway on a number of horses without any difficulty for the previous ten years it never occurred to me that Charlie would make an issue of it. But when he did, he had to be made to do what I told him to do. I said 'Please will you?' and he said 'No, I won't!' so he damn well had to. We would have gone through that gateway if I had had to stay all night. But of course at no time did I hit him – it became purely a battle of wills, which I had to win.

It is this battle of wills which is the essential thing with any horse. You must prove to him, without hitting him or ill-treating him in any way, that your will is stronger than his, and in a very short time, if you ask him to do something he will do it without much complaint or objection.

He will do this for a number of reasons. First, because you have established yourself within the herd hierarchy: he will see you in terms of the natural herd situation, placing you as the dominant horse to which he defers. Thus, you are making use of the training which he has undergone from birth, conditioning him to obey the dominant horse. Second, because by making his work enjoyable, you will have conditioned him to expect an enjoyable experience even after doing something he dislikes doing. Third, because by this time you should have created a bond of affection between you and the horse, he will want to please you. And last, he will have been conditioned to expect reward in the form of praise and being caressed. Thus conflict can, in most cases, be eliminated.

5: *Using Conditioning Theory: 2*

Every kind of learning known to us can be analysed satisfactorily into some combination of classical and operant conditioning; but, except for training by fear, we seldom use pure classical training in teaching a horse.

As I explained earlier, operant conditioning differs from classical conditioning in (a) the stimulus situation, (b) the response made, and (c) the relation to, and type of, the reinforcement following the response. I will now try to make these differences a little clearer. In operant conditioning the stimulus involves a whole situation, not a single brief event such as the ringing of a bell or flashing of a light, of classical conditioning. So the subject makes a variety of random responses: wanders around, looks at things, pushes them. It does not give a specific elected response as in classical conditioning. Finally one of these responses will bring a reward or avoid punishment. Put all this together and you can say that operant conditioning consists of learning to perform some random act (selected from a number of possible actions within a situation) which leads to a reward or avoidance of punishment.

Operant conditioning can be demonstrated experimentally with rats in an operant chamber. This chamber can be used in many ways. It may contain two or more levers, or two or more lights, a feeding place into which pellets can be dropped, a drinking place for water and a metal grid floor for applying electric shocks, and such a chamber allows one to study all sorts of learning. But reduced to its bare essentials it consists merely of a box with a lever protruding from one wall and a food cup below it for rewarding an animal, usually a rat, with food. Attached to the feeding lever through an electrical circuit is a device that makes a recording on paper each time the rat pushes the lever. This is called a 'Cumulative Recorder' because one response moves the pen one unit, another response another unit, a third response a third unit and so on, and the responses are cumulative, i.e. they add up. Since the paper moves at a constant speed, a steep line on the recorder means that the rat is making responses in quick succession; a flat line means that he is making few responses.

Suppose now that a hungry rat is put in an operant chamber which is hooked up to a cumulative recorder. Since the box is unfamiliar to the rat, and unfamiliar things tend to evoke fear in animals, the rat first shows signs of fear; but these signs soon fade as the box becomes more familiar. The rat starts to explore it and it does many things – it sniffs at the walls and crevices, paws at the walls and floor, stands on its hind legs and runs along the floor. Eventually, by chance, it depresses the lever, a pellet of food is released and falls into the food cup. There follows the click of the feeding mechanism and the sound of the falling food pellet. This is the rat's first correct response to the first reward in the chamber.

In the experiment we carried out with such an operant chamber, the period between the time the rat was placed in the box and the time it made its first reward response was just over a quarter of an hour – sixteen minutes, to be precise. Another minute passed before the rat noticed the food and ate it. The rat didn't learn anything from this first experience, but the food pellet, which the rat ate because it was hungry and aroused, caused it to explore with greater vigour. As luck would have it, the rat didn't strike the lever again for eighteen minutes, that is thirty four minutes after the experiment began. At forty-five minutes it made its third response and at sixty-two its fourth. At this point, the rat began to get the idea. It had, in fact, become conditioned, and responses started coming rapidly. From then on the rat alternately pushed the lever and ate the pellet as fast as it could.

This experiment illustrates the conditioning of an operant response, of the rat's operant behaviour. At the beginning of the session it was sniffing, pawing, running, standing and incidentally pressing the lever, but only one of the responses – pressing the lever – was rewarded. This response was one it learned after a few trials at pairing responses and reward. Note carefully that the rat was required to make the responses himself – it wasn't a reflex elicited by the experiment, as in classical conditioning.

You may be wondering what a rat in a wooden box in a laboratory has to do with training horses. The answer is that the process described above demonstrates the basic principles of all learning. In his meandering around the wooden box, the rat by chance touched the lever and then again by chance he touched it a second time. But the third, fourth and fifth time he touched it, he was doing it at shorter intervals because he had quickly learned which end of the cage the food was likely to be, and he was looking for more food. Seeking for food, he touched the lever; and by the time he had touched it half a dozen times, he was associating the lever with the food. Eventually he touched the lever whenever he was hungry because he knew that the lever would bring him food.

Now, this process is very closely relevant to training horses. One of the first things about learning that it demonstrates is the importance of placing the subject in a situation where he is *likely* to do the right thing – in this case, the lever near to the reward. So it is up to you, as the horse's trainer, to put your horse into the situation where he is most likely to do the right thing.

Take as an example the task of teaching a restless horse to stand still while you get on to him. If you try to do so in the middle of a field, he will be free to move around you and there is no possibility of him doing the right thing and getting praised (rewarded). You have instead increased the chances that he will do the wrong thing – knock you over, tread on your toes, which means that you are going to clobber him (because anyone who has fifteen hundredweight of horse standing on their big toe clobbers the nearest thing handy !). But if on the other hand you stand him in a corner, or, better still, a space that he must stand in square without moving, you can get on him and then praise him for standing still, so he will very quickly get the idea.

In the first of these two situations you have made training extremely difficult. In the second you have made retraining a badly behaved horse infinitely easier. Now, when he is stationary for you to get on, the next step is for you to get on him in a gateway, facing the corner of the gate. When you're on you open the gate from the horse and go on your way, and by steps from this, he will learn to stand still whenever you want to get on him.

Similarly, you can make things easier for yourself when teaching him to jump. If you take him out and put a three-foot fence, eight-foot wide, in the middle of a field and trot him into it, he will stop at the fence, go around it and do any of a number of very peculiar things. The one thing he won't do is jump the fence! But if on the other hand you take him down to a lane containing five or six poles, two feet or two feet six high, and let him follow half a dozen other horses, letting him watch them go one after another, then cantering him up the lane over the poles on the tail of the last horse, he will pop over the five or six poles without thinking about it. When you make a great fuss of him he will know that he had done the right thing – you have put him into the position where he is likely to do the right thing and get the reward.

This is operant training. In the second of these examples, the horse has had the stimulus of the excitement of cantering with other horses. He's also had the example of the other horses jumping and the stimulus of your keenness and enthusiasm, and his response to these stimuli has been rewarded. Unlike Pavlov's dogs who, when the bell rang, were fed regardless of whether or not they responded, your horse got rewarded only when he gave the correct response. This is the difference between operant and classical training.

It should by now be clear that the learning most common in everyday life is some form of operant training, though it is often combined with classical training. Here is a simple example of operant and classical learning in the natural behaviour of horses together. A group of horses in a field learn to follow the lead horse instinctively partly by classical training – because if they try to pass him he will clobber them. This may be observed when you see another horse approaching the lead horse, who will swing his head round as if to bite the offender. This has a direct link with old Papa Pavlov's dogs in boxes: instead of a bell being rung and the dogs salivating in anticipation of food, the signal is the leader's head being swung round and the response is the second horse jumping back in anticipation of a bite. But horses also learn to follow one another by operant training. Here, early training is the key, because if a foal doesn't follow its mother about it doesn't get fed, since the milk bar will have disappeared over the horizon. The foal is free to do a number of things, such as play with other foals, stand still or gallop around the countryside, but unless he follows his mother he loses the protection that his mother affords him and dies of starvation. So, operant training at the earliest stage in a horse's life teaches him to stay within a group for protection and feeding, while classical training teaches him his place in the herd. Both negative conditioning in the classical training, and positive conditioning in the operant training, are at work.

Operant conditioning, even in the horse's natural life, may be quite a complex affair. A large number of factors may be present, only one or two of which are relevant to the proper response. A perfect example of such conditioning was offered once by Chico and Spitfire. Chico was a two-year-old thoroughbred stallion, and Spitfire my old pet pony. To most people keeping a horse like Spitfire would be like keeping a lion as a pet, but to me her incredible ability, determination and her desire to excel more than made up for the fact that she was unpredictable, bad tempered and bloody-minded. I wanted to breed a really good foal from her, and since I had a nice thoroughbred two-year-old I decided that he would do the job perfectly.

For a few days I turned him out with Spitfire for an hour or so. At first she would have nothing to do with him, then she got used to him following her around while she was grazing. I knew that four or five days later she was due to come into season. Chico, when I let him out in the mornings, would canter across, screaming his head off, to where Spitfire was standing. She would put her ears back and tell him to stop fooling around, then wave a leg in his direction to tell him that if he didn't behave himself she would kick his teeth in. But, the day before she was due to come into season, when he came galloping across to her she didn't put her ears back to him and he stopped short in some surprise, rather suspecting that she was going to spring a trap, only waiting for him to come close enough for her to have a piece of his anatomy for breakfast. But she just stood there with her ears pricked, so he trotted round her with his head held high and his tail cocked up in the air two or three times, to show her what a handsome gentleman he was. She stood, moving slowly round watching him, and then he came over and sniffed her nose.

When Chico sniffed Spits' nose and got no adverse reaction, he took a step forward, arched his neck and sniffed again. She nibbled his shoulder and in return for this compliment he nibbled the back of her neck. She put her ears back and squealed, he took a step back and she took a step forward, so he nibbled her neck again and she squealed again, then he nibbled her shoulder and her front leg. This was taking too many liberties for Spits, so she bit a piece out of his backside and told him to shove off.

The following morning was a glorious May morning and you could see the mountain sprouting green shoots of grass, straight and firm like spears pointing at the sky, and the leaf buds of the laburnum trees which later crown the top of the mountain with flowers of gold were bursting out, revealing their delicate shade of green. When we got to the field Spits was waiting at the gate for Chico. I opened the gate and took him in and he very warily dodged to the other side of my body, just to make quite sure that he wasn't going to be in any trouble again. I turned him loose and he went belting round the field for a couple of minutes, then came back to say hello to Spits. She gave a low whicker which sounded very much like a sexy chuckle to me, and put her head out and sniffed Chico's nose gently. He sniffed back, walked forward and nibbled her neck. She showed her obvious pleasure by just licking his shoulder. Chico found this very pleasant, as no doubt Spitfire did, and with his teeth he caressed her neck, her back, her withers, her shoulders and down her front leg. Spits went down on one knee and squealed, Chico immediately drew back in alarm, Spits stood up and Chico nibbled her shoulders and her leg again. She took a step forward so that he was nibbling her side and then her hind leg. She turned her backside towards him, lifting her tail to one side as an invitation. Chico had no idea what was expected of him, so he leant forward over her to nibble her neck again, but since her backside was in the way he went half up on his hind legs so that he could nibble the place that he knew was acceptable, Spitfire stood still and Chico came forward.

When it was over, Spits shook herself and wandered off, but Chico, having learned that this was a pleasant occupation, pursued her and about half an hour later served her again, this time with very little fuss and very little loveplay beforehand.

This is a very simple example of how a combination of natural urges plus instruction, in this case coming from another horse, can teach a young and ignorant horse very quickly. Seen as an example of operant learning, Chico did a large number of things – he trotted in a circle, he sniffed Spitfire, he nibbled her, he walked round her – at random, but until he did exactly the right thing he got absolutely no physical satisfaction whatsoever. Spitfire, being an old and seasoned mare, of course knew exactly what she was doing and was able to help in teaching him; but even if she had been a green two-year-old filly herself, exactly the same thing would have happened, except that the two animals would have learned at the same time, so the process would have taken longer. Chico was used on a number of mares during the summer months, his son out of Spitfire being an absolute beauty, a liver chestnut with a silver mane and tail whom we immediately christened April Fire.

When a horse learns to make one response to one stimulus and another to a second stimulus, this is called discrimination learning. This kind of learning may be classical or operant. Pavlov taught his dogs to discriminate between a bell and a buzzer,

simply by reinforcing one with the giving of food and not reinforcing the other. The usual procedure was first to train a dog
to give a conditioned response, i.e. to produce saliva because
of the ring of a bell, then for the experimenter to sound a buzzer
without following it with food. At first when the buzzer was
sounded the dog salivated in response – by stimulus generalization, because the bell and buzzer sound somewhat similar. But
as trials with the buzzer went unrewarded, the dog salivated less
and less, even though it continued to salivate at the sound of the
bell. Eventually the dog would take no notice of the buzzer whatsoever, but always took notice of the bell. It had learned a
conditioned discrimination.

Operant situations can also be arranged to teach discrimination. An experimenter may wire an operant chamber to attach
different rewards to different stimuli, and the animal has to
learn by random response to distinguish the stimulus needed to
achieve a goal.

Discrimination learning is one of the most important aspects
of training a horse, because so much success in competitions, for
instance, depends on such skills as the ability to discriminate
between different tasks and obstacles, between dressage and
jumping, and between one kind of jumping and another. The
horse also has to discriminate between the moods of his rider,
the desire of his rider, the abilities of his rider and the signals
that he receives. And most important of all, he must discriminate between correct and incorrect behaviour in an emergency.
If you are in tune with your horse, he will very quickly learn
to discriminate between being correctly placed at a fence and
coming into a fence all wrong, and if he is wrong to correct
himself, and put in that extra effort required of him.

When you are indulging in competition, hunting and indeed
anything other than general hacking, this fine discrimination of
response is extremely important. Just as, when I am talking to
people about learning, communication or psychology, I speak
and behave in a completely different manner from when I am
in a jockey's tent at a point to point, so horses must learn
appropriate behaviour to different circumstances. A horse

with a fine sense of occasion was Clear Reason by High Treason. I had a string of two or three horses going at that time, and there was no question whatsoever that Clear Reason was by far the worst of them. When the others had finished the gallop, he was trailing a hundred or a hundred and fifty yards behind. When I'm training I like to train against the stop watch, and our best horse at that time was doing the two-mile gallop, which is a very stiff and difficult one, in four minutes and eight seconds. The best that Clear Reason ever did was eight and a half minutes on the gallop. Come the first race of the season though, since I'd spent a small fortune on feeding the lazy beggar, I pushed him into the lorry with the other horses. When we got there we sorted them out, and a very good maiden we had was put in the maiden's race, my best horse Argonaut was put in the open, which only left the hunt race for Clear Reason.

Clear Reason's race was the third. I had already finished second on Argonaut, and felt very unlucky not to win. Came Clear Reason's race and I was expecting rather an uncomfortable bump round, pulling up half way round the second circuit because the rest of the field had already passed the winning post. We cantered down to the start. He was a little slow starting and went into the first fence last, but he took off well back and he went on improving until half way round the second circuit of eight, when he was lying third. I just couldn't believe it, he still didn't seem to be tiring, so I tucked him in behind the two leaders and he floated round without any difficulty. Towards the end he tired because he wasn't very fit, but I finished an easy third.

I ran him twice more, doing extremely well with him and then I had an extremely good offer for him and sold him. But I always thought of him as a fine example of discrimination learning: at home he could see no reason for bothering, but on a racecourse he could see every reason for getting down and galloping. I don't think I ever had him completely tired, but since he was only a young five-year-old I never really pushed him out.

In everyday life of course a horse has to discriminate amongst ordinary objects. We are in a low flying area, which means that the air force fly over us at about fifty feet with young fools of pilots dive-bombing us, breaking the sound barrier and generally doing their best to make an infernal nuisance of themselves. Horses, to start off with, are extremely alarmed by this and tear round the field in a panic, but after it has happened two or three times, the sound of an aircraft wooshing down over their heads doesn't worry them at all, and after a month they won't even lift their heads from grazing. They have learned to discriminate between the noise of a car wooshing past, which does frighten them, and an aeroplane coming down from a great height and making considerably more noise. They have learned that an aeroplane cannot hurt them and a motor car can. In actual fact we then use their acceptance of the low flying tactics of the idiot in the cockpit to train our horses to behave quietly in traffic.

For demonstration purposes I have taken a thirteen-two pony around our cross country course without a bridle or a saddle simply by shifting my weight to direct him to the next fence – leaning forward when I wanted to go faster and leaning back when I wanted him to steady. I started that piece of training quite simply by using a bridle and a saddle but exaggerating my weight positions until I reached such a point that his discriminatory learning had taught him that the positioning of my body indicated the speed that he had to go, and what he had to do. Also, when I have finished schooling a racehorse, which is usually after the eighth or tenth time I have ridden him racing, the horse will place himself at the fence precisely – he has learnt the exact distance that he needs to take off from each fence, the difference between an open ditch and a plain fence, a plain fence and a water jump, so that he will automatically shorten or lengthen his stride to jump the fence.

A very interesting incident happened in the race on Fanny at Aintree that I describe in detail at the end of this book. Before going to Aintree I had raced her in three hunter chases. Until she tried hunter chasing, she hadn't seen a water jump, but being a very clever jumper she learned to jump the water without any difficulty whatsoever; she also learned that water jumps were twelve feet wide. When we came to Aintree she jumped the whole course precisely, except the water jump. She made no mistake whatsoever, in fact when she jumped Beecher's there were two horses on the ground, one straight in front of where she was due to land and the other slightly at an angle. She jumped, changed legs, popped over the horse lying on the ground, changed legs again and went round the second horse who was down, damn near getting me out of the saddle in the process. But the water jump was a different question altogether. She knew from her previous learning that she had to jump twelve feet – so she jumped twelve feet, dropping her hind feet in. She hadn't had a chance to discriminate between the bigger water jump at Aintree and the small water jumps she had been accustomed to jumping, and I've no doubt whatsoever that if I'd taken her round Aintree again she'd have made no mistake whatsoever at the water jump.

Motor learning, sometimes called psycho-motor learning or skilled learning, is learning how to do something well. The emphasis in motor learning is on learning *how* a particular response is executed. And motor learning in horses involves the co-ordination of responses and skills so that the movement gets as near as it can to perfection.

All forms of jumping and training involve a certain amount of motor learning, so this is one of the important branches of learning in developing the finer points of horsemanship. Development of skills can be the most important and involved side of training a horse, and a large number of books have been written by a large number of people about how this should be done. In actual fact the basic principle is extremely simple.

When we described teaching a horse to jump earlier, we suggested you canter a horse up a lane behind four or five others and pop him over some low fences, which he enjoys doing. When you follow this up over a period of a week to ten days doing it every day, before very long the horse will be dancing about with excitement as you approach the jumping lane, because he is associating the lane with (a) the pleasure of jumping with other horses and (b) the new experience of taking off from the ground and jumping fences. The sense of achievement when he has done something he had considered impossible, gives him a great deal of satisfaction. And the praise he gets from you is an added stimulus. All this adds up to him connecting jumping with pleasure. At the end of the week he will have discovered that if he hits the fence he hurts himself, so he will begin to clear the fences well, often by an exaggerated two or three feet.

In theory, if you progress from here, raising the fences by degrees, he will continue to jump and enjoy jumping until you get to the height that marks the limit of his physical capabilities. But unfortunately, chance, to a small extent, and human incompetence to a very large extent, stops most horses jumping when they get beyond about three feet. What happens is that, by chance, the horse falls: he slips when he takes off badly, and hurts himself, so he stops jumping. Pure laziness will also stop a few horses jumping.

It is with the remainder, the horses who do go on to jump to the limit of their capabilities, that human incompetence comes in. The idiot on his back – and that includes myself, of course – tries to make him jump a fence that is too big for him, or too big a spread, before he is ready. He jogs his mouth, hits him when he is doing his best or when he had made an innocent mistake. All these things conspire to make a horse nervous when jumping, which will start him refusing his fences and so stop his natural progress. This means that you have to go back to a much lower height and eventually you reach a limit of about three foot or three foot six which is the limit of what most horses and ponies can jump easily, even though their natural ability makes it possible for them to jump five or six feet. This limit of three foot to three foot six is usually due to the competence level – or incompetence level – of the rider.

We had an example of this problem with Biddy last summer. I know a fair bit about riding cross-country, having ridden cross-country all my life, but since I have started one-day eventing I have to do a certain amount of show jumping at which I am a complete ignoramus, since jumping cross-country and show jumping are two completely different sciences, both of which have to be learned from a very early age. However, it seemed to me at the beginning that riding show jumping fences was only riding cross-country fences that knocked down.

I very soon learned the error of my thinking, because show jumping, it appears, is all about related distances between your fences. You have to take them on a short, even stride, whereas when we are riding across-country we take the fences as they come, and put in a short stride where necessary to get over. On an extra big fence, either a short stride or an extra big jump gets you over the obstacle. At this, Biddy and I were doing extremely well, completing our novice cross country easily and well within the time limit, and when we took on the crack riders in open competition, we put in extremely polished performances, usually doing the fastest time of the day.

Show jumping, however, was a different cup of tea altogether. We left a large number of coloured jumps demolished from the tip of Pembrokeshire to the North of England, so both of us had to go back to school and learn something completely new.

This is just one typical example of how a fool on a horse's back can impede the performance of the inoffensive animal between his legs. But the principle of training a horse to do anything to perfection is to take it slowly and gradually, operating well within the horse's capabilities, giving variety and the feeling of achievement and having done well. At the end of each round, even if the horse makes a mistake, you go back and jump two or three much smaller fences, so that you finish up with that sense of achievement again. This applies to any form of training and any form of riding.

This sense of achievement is very difficult to attain in the more precise schooling movements. It is easy enough when teaching anything to a horse in the early stages to give him the feeling that he has done something extremely difficult. Similarly, when you're jumping a slightly larger or different fence, the sense of achieving the impossible provides him with the motive

to jump something even more difficult. But when you're teaching the horse very precise movements, this is much more
difficult.

Suppose you are teaching a horse to stand completely stationary. Since to most horses the release from the stationary position
will be a reward, he will tend to anticipate the movement, in
order to attain the reward more quickly. Similarly, teaching a
horse to go from a collected walk into an extended one is complicated by the fact that he will try to trot, there being no
obvious incentive to walk instead, other than the restriction of
the bit. You will have to deal with this problem first by creating
a situation which discourages him from trotting. Part of this
preparation will be in the way you feed the horse – on molasses,
boiled barley, flaked maize and a little oats for a show horse,
instead of racehouse nuts or a main diet of oats. You can certainly plan the training so that it is followed by an immediate
reward – for instance, dinner. But the difficulty still remains
within the training itself. One remedy is to have an audience
watching you when you are doing these precise movements, who
applaud when the movement is done correctly and express
derision when it is anticipated or performed badly.

Improvement may also follow if you produce the horse
groomed, plaited and perfectly turned out when planning
dressage training, then take the plaits out when the dressage
training is finished. Or a method which is effective with some
horses is to have a tape recorder playing music timed to coin-
cide with the horse's movements. Punk rock and modern music
do not work, and in my experience the Death March with a
good drum beat, speeded up to your horse's natural movements,
is best for both the collected and the extended walk.

These are just some ways you can improve your schooling of
precise movements, so that apart from your own vocal and
caressing rewards the learning is made easier, and a sense of
achievement for doing each movement precisely (which is in
itself a reward) maintained.

To put it concisely, the secret of motor learning, like that
of other learning, is implanting the motive – in this case, to

attain perfection. This is essential in teaching a horse to jump a show jumping fence precisely, or in achieving the perfect dressage movement. The reinforcement and reward in motor learning is the precision and perfection of the movement, not the action or the movement itself.

6: *Fear Conditioning and Avoidance*

Avoiding something – like a jump – may be a bad habit in a horse. It is an essential skill, which, like other behaviours, all animals must learn, but it is also something that we may have to get them to unlearn.

Learning to avoid another horse, object or situation is a combination of classical and operant learning, linked in a special way, and depending on something known as avoidance motivation. That is, before avoidance conditioning can occur, fear and escape conditioning must have taken place.

Fear conditioning takes place when some neutral stimulus is paired with a stimulus which naturally causes fear. I can illustrate this with Swallow, a nice palamino foal I bought a year ago as a Christmas present for my daughter. The horse I was hunting at the time was a big sixteen-hand skewball called Kerry, and he for some reason took a violent dislike to Swallow. Every time Kerry passed Swallow's stable door he would make a dive at him, and Swallow became so terrified of the big horse that he had only to hear Kerry's stable door being opened and he rushed to the back of his loose box and stood crying there until he was sure the horse had gone out of sight. He was pairing a neutral stimulus – the sound of a door opening – with something that he was afraid of, i.e. Kerry, so that the sound of the door opening would terrify Swallow even if Kerry wasn't coming out of it.

I dealt with this problem by shifting Kerry into a different box with a different sounding door – Kerry's own stable door used to stick so you had to give it a thump when you opened it, and the sound of the thump had become the signal that terrified Swallow. For two days Swallow still showed signs of terror at the sound of the empty loose box being opened, but by the fifth day he was taking no notice of it at all. But I put Kerry back

54

in, and immediately Swallow's fear reappeared at the sound of the door opening.

The pairing of Kerry and the sound of the door had, in good Pavlovian fashion, produced a conditioned fear in Swallow. Once fear had been conditioned and Kerry's aggressiveness had been paired to the fear-producing situation, Swallow's reaction was to escape from it. This led to Swallow desiring to escape before Kerry appeared; his natural intelligence told him that Kerry was coming when his stable door was opened, so he became afraid of the sound of the stable door.

Now this tendency to be conditioned, to associate a neutral with a fear-producing experience can be of great use, when training horses. Most people for instance can induce fear in their horse simply by scolding him. This works because the horse has learned to associate the scolding with punishment, so, provided that the scolding is reinforced from time to time with punishment, the scolding alone will usually be sufficient. The noise you make scolding a horse does not hurt him, of course, but the horse associates it with a painful situation and takes evading action.

The evading action is the result of another conditioning – escape conditioning – which may have been produced by either an unconditioned stimulus or a conditioned stimulus. Escape conditioning is a special form of operant learning, as illustrated by the following experiment with dogs.

A dog was placed in a box which was divided into two halves by a low barrier. A bell was sounded and, half a second later, a shock was given, passed through the floor on one side of the box in which the dog had been placed. The dog had to jump the fence dividing the box into the other half to escape from the shock. By the third trial the dog jumped the barrier *before* any shock was administered. Thereafter, he jumped the barrier on each occasion the bell sounded. Thus to begin with, the dog jumped the fence to escape from the shock, but once it was answering to the bell it was avoiding any shock whatsoever, and not just escaping from it. The bell was arousing a fear and to avoid the fear the dog was jumping the fence, even when there was no electric shock applied.

One of the methods we use to teach a difficult horse to load in a lorry or a trailer makes use of this method of learning. We use this only if our preferred techniques – coaxing him in with a bucket of nuts, and leading him up to the trailer and lifting his feet one after another up the tail board – fail. First we make a chute with a couple of gates opening to the back of the trailer or lorry, then we lead the horse into the chute. I stand behind him with a handful of pebbles, cursing him at the same time as throwing a pebble at his bottom and I keep this up until he decides to avoid the perpetual stinging of his bottom by going into the lorry. Immediately we make a fuss of him, then take him out of the lorry and repeat the process. We do this over and over until I have only to curse him once and he dashes into the lorry like a shot rabbit.

Fear is an extremely important tool in your handling of the horse, but it is a very finely edged one. On the other hand, the horse must respect you, and a horse will not respect any animal that is below it in the social hierarchy. In a herd the horse respects the most aggressive animal, which all the other horses have to respect, so your horse must equally respect you. But it is a completely different matter if he is afraid of you. If he is afraid of you as a person, everything you are doing in his training will be undermined, and you will never get him to want to do what you want him to do. What he must fear is the fact that you will punish him if he does wrong. It is rather like the position of father and son : a son must not be afraid of his father, but he must know that if he steals the apples off his father's favourite apple tree he will have the living daylights whaled out of him. In that case he fears the punishment for theft, though he doesn't fear his father. Similarly the horse must fear the punishment for doing wrong, but must not fear the person administering it.

A friend of mine has a vile temper and on one occasion when I was riding with him, his horse shied violently and bucked. and my friend came off. I caught the horse and came back and, to my horror my friend proceeded to whale the living daylights out of the animal. I was just going to get off and, in temper myself, stop him with my fists, when he stopped hitting him. His horse walked up to him and just rubbed his head on his shoulder. The horse knew when he was caught that he was going to get a hiding; even so, when he had had the hiding, he showed no fear whatsoever of his owner – in fact great affection. The horse knew he had done wrong in bucking, accepted his punishment and that was it – over and done with. Whilst I don't like to see a horse receive a hiding, the reaction of the horse to his owner was proof of the affection between them.

There must always be this bond of affection between horse and rider so that the horse will do things simply to please the owner. He must enjoy his work – in training this is an extremely important thing. Dull and repetitive work produces a dull and bored horse. At the same time the horse must be obedient; by this I don't mean the obedience of the stern German method of horse training, but the obedience that comes from the horse's knowledge that if you tell him to do so and so, then he must do it; so he does it because it is easier to do it straight away than after a long argument, and because he has a great desire to please you.

My work with Charlie shows how fear conditioning and affection can work together, and not against each other, in horse handling. He soon had a certain amount of affection for me, though the bond of affection between us was not yet very strong. But after riding him for only a few days, Charlie knew that if I asked him to do something he had to do it. The first time I asked him to do something when he didn't want to there was an argument that took well over an hour. The next time the same sort of situation arose, the argument lasted only a minute or so because Charlie already knew, after being handled firmly but kindly for a few days, that he was going to have to do as he was told anyway. He next reached the stage where he was really enjoying his work and he saw the reason for doing what he had considered previously to be quite impossible.

The power of fear can be such that a comparatively trivial event can interact with later experiences until the fear governs a whole range of behaviour. It becomes a conditioned response to other stimuli connected with the original situation. Something quite insignificant may happen to a foal which will multiply itself by this process of association until, when you come to gentle him, the fear has grown out of all proportion. I remember a foal that was in a field where there happened to be a great number of burrs. These got caught in the foal's coat and inside his legs, so each time his owner caught him, to keep him tidy and stop the burrs irritating him, he would pull them out of the foal's coat. But of course he couldn't help pulling a little hair with each one, and this was slightly painful to the foal.

This went on about once a month right through the summer. By the autumn the very sight of anyone approaching him petrified that foal, even though his mother was quite quiet and easy to handle. So whenever he was caught a considerable struggle ensued, which only frightened the foal more. By the time I got him as a three-year-old to gentle, he was conditioned to terror of man.

Another horse I rode had been kept throughout the winter in a stable with a very narrow doorway, and whenever he came out or in, he banged his sides. So he got into the habit of going through the door in a rush. By the spring, if you opened even a twelve-foot gate for him he would go through it absolutely flat-out for fear of catching himself on either side. On one occasion he caught my knee on the gatepost, knocked me off the saddle and I got up limping on one leg with a sore backside.

The increase in fear in cases like this may be so gradual that you don't notice it at the time, but it can multiply itself until the horse is in a state of near terror. And by that time it is a problem to both horse and trainer.

Just to show you that pleasure can be a pretty powerful conditioner too, and that for your own good you shouldn't be afraid to balance things out with a little fear conditioning when necessary, let me tell you a little tale about Chico. When he was losing his coat in the spring Chico used to back up to a tree to scratch his backside, which he enjoyed doing very much indeed. He had in his paddock a favourite small tree with a low branch which scratched his back as well, and this tree would be seen shaking about all over the place while he scratched himself on it. While I was riding him idly about the field one day, we wandered over by that tree and the next thing I knew I had a branch up my backside and I was sitting half way up on his neck whilst he proceeded to scratch himself. Even though at that time he was being groomed every day and he had no excuse for an itch at all, he'd been conditioned during the spring to associate the tree with pleasure. The fact that I was sitting in the saddle, admittedly half asleep, made no difference to him. He went to his favourite tree to scratch, boosting me out of the saddle in the process.

When he discovered what he had achieved, he found an opportunity too good to be missed. Forsaking the pleasure of scratching his backside, he took full advantage of having me half way up his neck to put in two colossal bucks, depositing me neatly on a bull thistle. Arching his neck and sticking his silver tail out like a banner, he floated round the field on his toes, each strutting stride showing his satisfaction at having got the better of the old man. Painfully I got to my feet, dropped my trousers, and removed most of the thistle spikes from my bottom. Then, replacing my trousers, I caught Chico who, seeing the unaccustomed sight of me standing there without any trousers, had come over to investigate. I got back on him and allowed him to wander back to the tree. As he backed into position, being well prepared this time, I caught him two hard cuts with my stick across his backside, which, being an intelligent horse, quickly taught him that scratching his posterior when I was in the saddle was not a pleasurable occupation.

So far I have been dealing with what happens when the simple conditions for learning exist. The conditions themselves, however, can be altered in many ways to make a difference to how effective the learning is, and to how well it is retained. Learning in the horse may be of two main kinds, (a) extending the horse's natural abilities and (b) learning a new task, or straightforward learning.

Now it's no damn good teaching a horse to do something

unless he remembers what you have taught him. For example, it is no good teaching a horse one day to stop when you say 'whoa' if he doesn't remember to stop when you say 'whoa' next day. And in memory retention fear again plays a significant part. Memory retention may in fact be governed by one or more of the following factors: fear; pleasure; habit; and subconscious memory. The longest lasting form of memory retention is based on both subconscious memory and fear. Something that has frightened the horse enough to have become part of his subconscious memory will be remembered long after he has forgotten something he has enjoyed or something he has done by habit.

We had an example of the power of subconscious fear retention in Cork Beg, my wife's favourite hunter fifteen years ago. She jumped a fence after hounds one day, cantered across a field and at the last second the horse saw the very thin wire of an electric fence in front of him. He attempted to jump it, getting his front feet over, but the electric wire caught underneath his tummy. Unfortunately the wire was live, so he continued across the field with the electric wire caught on his stifle against his most tender spot, getting a shock every two or three seconds. He bucked like hell to get rid of the irritation, and deposited my wife in two feet of mud before he finally got rid of the electric wire. No permanent physical harm was done, but from then on it was literally impossible to get him to go anywhere near electric wire. You could put a piece of wire straight across the yard and put his feed two feet on the near side of it, so that he had only to go within two feet of the electric wire to get his feed; and he would go to within four or five feet and stretch his head out trying to reach it. But he would starve rather than go any nearer, and he retained this memory for over fifteen years. Until he died at the age of twenty-two, we could shut him in a stable or field simply by tying a piece of string across the gateway. And he remained one of the very few horses we failed to teach to jump barbed wire, simply because we could never get him near enough to the wire to jump.

The memory of pleasure is much shorter lived, though how much shorter is of course dependent on the degree of pleasure, and a very intense pleasure may indeed be remembered longer than a mild fear.

The behaviour of a mare we had at home called Darling is a good illustration of memory retention based on pleasure. She was bred from as a three-year-old, and she enjoyed muzzling her foal so much that for ever after, whenever she was turned out with mares and foals, she would try to steal one of the foals. We used her a lot for training young horses – she would take a yearling, two-year-old or three-year-old and treat it like a foal. Now, this passion could not be seen in terms of pure mother instinct, because other mares do not behave in this way. But Darling always had her 'baby' with her, even though the baby was sometimes two or three hands bigger than she. She looked after it and taught it to follow her as a foal would, simply and solely because mothering a foal had given her such pleasure that she sought to repeat the pleasure by looking after another immature animal. She got so attached to these substitute babies that when, at ten or eleven years old, she had her second foal, she rejected it for the first four of five hours, until we had removed the two-year-old that she was mothering at the time.

Habit retention is seen everywhere – for example your horse carries you round largely from habit, because he is used to doing it. He will form bad habits and good habits, often quite by chance. It often seems that bad habits are a damn sight more difficult to get rid of than good ones and good ones far more difficult to establish than bad, but in actual fact this is not so. The depth of, that is the length of time that the horse retains, the habit is the only real measure of the difficulty of curing it; and it seems more difficult to get rid of bad habits than good ones because most of us have allowed bad habits to develop naturally from our own bad riding. To cure these we have to improve our riding to such a standard that the habit is eliminated automatically, and this entails learning something ourselves. Since the thickness of the human skull makes the base of

Mount Everest resemble an ice cream wafer, this is a task few of us approach easily.

For instance, if to cure a vice in a horse the rider has to learn to ride with a slacker rein, while he is concentrating he may do so, so curing the fault in the horse. But when he relaxes he will subconsciously tighten his reins, thus making his horse revert to his bad habit.

The concept of subconscious memory retention, already referred to in relation to Cork Beg and his long-remembered fear of wire, applies to any continued reaction by a horse (or other animal) to an event which has itself been forgotten. Another example might be the behaviour of a mare we had some years back. We took her to the stallion and she bred a foal from him. We didn't breed from her the second year; but in the third year, when she was in season, she got out of her field and went two miles to the field where she had previously been served by the stallion (and incidentally got in with him and I had a free service, but that's another story altogether! The argument about whether I should pay for the service is still going on ten years later!). In this case the mare's memory was re-awakened by the physical changes in her body, which took her over the fence and along the road two miles, over another fence and down to where the stallion was.

So if you want your horse to remember what you are teaching him, you need to bring as many of the factors in memory retention as possible into play; but again, fear has to be treated with great care. It is an intense emotion, it can lead to avoidance, and a behaviour instilled by fear is exceptionally difficult to eradicate.

Another of the important conditions for learning is the degree of arousal – or emotional excitement – appropriate to the task. For example, when getting on a horse for the first time you want the horse to be relaxed, happy and unafraid. This means that you want a low state of arousal. On the other hand, when teaching him to jump, you may want a high state of arousal. The degree of arousal required also of course depends on the temperament of the horse; a lazy or timid horse may require a very high state of arousal to make him jump, an alert and confident horse a comparatively low one.

Arousal may be affected by several things. First, by feeding, because a horse that is starved with a staring coat and pot belly will be extremely difficult to arouse, and hence difficult to teach. But a horse that has been shut in a loose box and fed like a fighting cock will be impossible to teach, since, after he has burst through the stable door like a bullet, he will be so excited that even if he doesn't buck you off into the dung heap, he will be dancing around so much that he will be able to pay no attention to the task. So the first thing to consider is how to adapt the feeding of your horse to what you want to teach him. (This is one of the things that makes combined training so difficult, since you want him quiet and restrained enough to do the dressage and show jumping, and yet excited, fit and bold enough to complete the cross country within the time limit.)

Arousal is also affected by the situation of the horse. A string of horses walking along half asleep will tend to quieten him down, whilst the same horses galloping will make him excited and lively. Fear, stimulated by a good belt across the backside, will wake him up and turn a lethargic plug into something that is worth riding. On the other hand, the relaxation of the rider and a caressing hand and voice will steady a horse considerably.

And this brings us to the mental state of the rider which is also a factor. If the rider is a mooning young woman dreaming of marrying Prince Charles in Westminster Abbey, her horse will be relaxed, taking a mouthful of grass now and then and thinking of fresh green grass. If, on the other hand, she is petrified and wondering whether the horse is going to buck her off in front of the next lorry, she will arouse such fear and excitement in him that he probably will dump her at the next convenient spot and return home for lunch, having learned nothing other than that the quickest way to get home to dinner is to deposit the rider as soon as possible.

Normally the horse should be aroused, but not too aroused. Up to a certain point, the higher the arousal level, the better the learning. Beyond that point the horse becomes too highly excited and most forms of learning are retarded. This can actually be illustrated experimentally on a graph, showing the general relationship between arousal and learning as an inverted U-curve : learning improves as the arousal of the horse increases to a fairly high level, but beyond a certain level it deteriorates. When high arousal fades into great anxiety or emotional excitement, learning is severely hampered.

Hence, if an experimentor is training a dog or a rat to jump a low barrier in response to an electric shock, he must be careful first to make the shock strong enough to make the animal react, but then not to make it so strong that it is overcome by its emotional state and bounces around the cage without paying attention to the warning stimulus. This excessive arousal can slow learning so much that in many cases the animal will not learn at all.

A further condition that affects learning is the motivation of the horse. The relationship between arousal and motivation is circular. It is after all obvious that an aroused horse is easily motivated, and a horse insufficiently aroused will be very hard to motivate. But the horse needs to be aroused to exactly the right level if he is to show the highest response to motivation. If your horse is grazing and relaxed and you rattle a bucket and shout for him, he will come over, motivated by the bucket. Younger horses and those most difficult to catch may be motivated by the example of other horses. On the other hand, the same group of horses galloping around the field in high excitement will ignore you completely – you can shout until you're hoarse and rattle the bucket until your arms drop off. That is how the relationship between state of arousal and motivation or response works : your horse, when asleep, has no motivation to jump whatsoever; raise the level of arousal and he is keen and alert and can respond to your instructions; raise the level of arousal further so that he is wild and excited and he will demolish the fence. So, the right level of arousal is essential. It is necessary that the horse should be awake and alert to the exact level that will bring the best response to any encouragement that you may give him.

What we are setting out to do is completely to change the behaviour patterns of our horse. The natural behaviour of the horse might be summarised, rather crudely, as : to avoid danger, to eat, drink, sleep and, if the weather is bad, to find shelter. But we aim to change that behaviour pattern to such an extent that the horse will gallop three miles and jump eighteen fences in a steeplechase, to jump unnaturally high objects in the puissance competition, to perform equally unnatural antics in dressage classes and in general to submit his will to that of whichever competent or incompetent human being he may be asked to carry hunting, riding or hacking around the countryside. By training, we not only get the horse to submit to these strange activities, but actually to enjoy them. We completely alter the horse's behaviour, and we do it by conditioning. When he does what we want he gets rewarded, and by repeating the reward – the praise or whatever – each time he does what we want, the horse will come to want to please us. First, pleasing us leads to a quiet life for him, without arguments; second it produces titbits and affection. This process, of constant conditioning by reward (and some punishment), is the crux of good training.

Training of course starts with the horse's natural responses, and builds on them. It means strengthening some responses (and weakening others) by a programme of regular reinforcement.

7: *Principles of Reinforcement*

Reinforcement is a general word, covering both reward and punishment, and it is anything which strengthens a response. It reinforces the response, or in other words it promotes learning. A reward is a positive reinforcement but punishment is a negative reinforcement. In either case the reinforcement is the road to learning. If it isn't used, or isn't used in the right way, no learning will take place. The horse is an extremely pleasant and kind animal who usually wants to learn. But if the thick-headed biped, who is endeavouring to impose strange and incomprehensible practices on the horse, relies only on garbled secondhand knowledge, weakness, ignorance or brutality as his training tools, the horse will learn very little. (It is as brutal to spoil, overfeed and then ask the horse to do something that is against its natural instincts without first conditioning it to want to do that thing, as it is to starve and beat a horse into submission.)

A horse is motivated to learn partly by his own natural curiosity, partly by his natural need for activity, and partly by the fact that he wants to please his owner. It is the last response that we work hardest to reinforce: and that is why positive reinforcement, that is the encouragement of the horse to learn for a reward, is so important, because this increases his desire to learn, since it leads to something pleasant. It increases his desire to please his owner since his affection for his owner is greater and the more he learns, the more he pleases his owner. In other words, by making sure the work is enjoyable you introduce another reinforcement. In learning as many as five or six different reinforcements may be used for the same piece of learning.

On the other hand bad horsemanship, ignorance and weakness may induce a revulsion against a particular piece of learning; thus undermining both positive and negative reinforcement.

66

Jasper was an example of a conflict of this kind. Whilst a sweet, kind horse normally, he had developed a revulsion against the tedium of exercise, which was necessary to get him fit for racing. His avoidance technique, which had been allowed to develop through weak handling, was quite simple: as soon as you rode him out of the front gate he would go straight up on his hind legs, swing round and shoot back the way he had come. His desire to please was in conflict with his wish to get back to his nice warm stable. The problem was to reinforce his desire to do what we wanted him to do. This we quickly achieved by determination, patience and ending his exercise by allowing him to gallop flat out up the jumping lane with another horse. This he enjoyed, and it also cured his main vice which was refusing to jump when racing.

The more the horse enjoys learning, the more he will want to learn, whereas negative reinforcement, i.e. punishing the horse when he has done wrong, reduces his desire to learn, since learning becomes an unpleasant experience. Negative reinforcement creates its own spiral – if learning is unpleasant, the horse will avoid getting into a situation where he has got to learn, and since ill-treatment produces a revulsion for his owner, his desire to please his owner will be lessened.

I can best illustrate what I mean by enjoyment in learning, by describing one of my favourite rides. On this ride we go up to the mountain through the forest, riding through the green avenues of larch standing in neat squares, precisely spaced line after line like a squad of guards on parade, stretching twenty or thirty feet above us, their tips swaying like feather dusters brushing the clouds. We turn down through the trees, while the branches slash our faces, punishing us for invading their lines, through to a ride at the bottom where with a plunge of excitement the horses bound away over three or four fallen tree trunks ranging from eighteen inches to two feet six inches high. Then we turn right down a steep track to the river at the bottom. The horses slide the last ten yards, landing with a splash in the pool, causing a cascade of water to sparkle in the sunlight. They stop and drink, standing in the water, some up to their knees, some up to their bellies. Then the first one raises a hoof and brings it down hard into the water, soaking himself, his companions and their riders. Soon they are all playing at it until I, who have got out of the way as soon as my horse has had a drink, have decided that the whole party has got wet enough for one day.

The consequence of this game is that jumping in and out of water is, to my horses, fun and I never have any trouble jumping into rivers or lakes out hunting or at cross country. But suppose, on the other hand, I had jumped each of those same horses into water for the first time alone, and a battle had ensued, they might have disliked jumping into or over water for ever after.

This preference for positive reinforcement does not mean that a horse should be allowed to get away with wilful misbehaviour. When a horse has bitten your backside he must be punished immediately. But if on the other hand the horse accidentally steps on your foot, even though for you the pain of one equals the pain of the other, it is your own fault for getting your foot in the way, and, far from punishing him, you should make quite sure that your boot has in no way damaged the horse's hoof.

The principle is quite easy to follow. Positive reinforcement should be used to encourage a horse to learn. Negative reinforcement is used to stop him doing wrong.

When Blodwen came to us she had a reputation for kicking – and she did kick badly. After she had taken a swipe at me the first time, I put on my heaviest boots and went back up to her. As I approached her and she took another swipe at me, I dodged, then brought my foot back as hard as I could and let her have one back hard in the belly. The message got home immediately. After that I could walk in and out of her stable and she never kicked me, and very soon she was absolutely safe for anybody to approach from behind. This story clearly illustrates negative reinforcement used in the correct time and place.

About twelve months later, Blodwen was eating her dinner when Pudding escaped. Pudding, being an extremely greedy pony, tried to steal some of Blodwen's food. I saw this happening and, not wanting a battle in the stable, went to get Pudding out. Pudding saw a kick from Blodwen coming, but being rather quick on his feet, dodged to one side and I received Blodwen's well-shod hoof in my posterior, which was extremely painful. But I was, in this case, just the unfortunate bystander. Blodwen had every right to kick at Pudding for trying to steal her dinner, and Pudding, being no fool, had naturally dodged out of the way, so the kick was just my bad luck, and certainly not a case for punishing Blodwen. It is a perfect example of where negative reinforcement should not be applied.

One of the most important things to remember about negative reinforcement is that after a battle, when the horse has done what it has been asked to, a great deal of praise and reward should be given. This is so that you can replace the negative with a positive reinforcement as soon as possible, and not be forced to repeat the punishment in similar situations in future. For similar reasons, if it should be necessary to punch or beat your horse, you should make sure you swear at him at the same time as chastising him. This is so that he will link the cursing with the clout, and after a while you may be able to use the cursing alone, without the accompaniment of the clout, as punishment. After chastising him, you should go away and leave him, to give him a chance to think about it and settle down. (This also gives you the opportunity for any surgery necessary on yourself). Then, after ten or fifteen minutes, you should return and make a great fuss of the horse, so that he does not associate you with ill-treatment only.

I had a very difficult horse to gentle, called Exeter Lady. She was just awkward, lazy and bloody-minded. Eventually I got her so that I had only to curse her and she would behave instantly. She started racing the following spring, but although several jockeys rode her, none of them got beyond the second or third fence, in spite of the fact that they were waving their whips like windmills. Eventually the owner had to eat humble pie and asked me to ride her. Instead of hitting her I swore at her going into the first fence, repeating the curses at each subsequent fence. I led from the third fence, using every expletive in my vocabulary. When coming up the straight, with the favourite up on my shoulder, my language became extraordinary. Yet no time did I touch her with the whip, since I knew that her reaction would have been to stop and start bucking. The outcome of this adventure was that the stewards warned me about my language in future races, but this was more than compensated for by a winning wager at 33 to 1.

8: *Teaching By Positive and Negative Reinforcement*

Positive reinforcement, i.e. reward and praise, should be used, but not be used to excess. Biddy gets no praise whatsoever when she jumps a three-foot-six fence, since it requires no great effort on her part. But on the other hand, when Pudding jumps three foot, he is jumping to the limit of his capability. He had to put considerable effort into it and he gets the blessings of heaven showered upon him. If I praised Biddy every time she jumped a three-foot-six fence I would have no praise left to give her when she jumped four foot six or five foot. If your horse automatically comes to you and is easy to catch, a slight pat or a friendly greeting is sufficient reward; but a horse that is very difficult to catch should be praised when he lets you catch him for the first time, and made such a fuss of that he will want to be caught again. So, it must be remembered that the reward should correspond to the effort involved.

Reinforcement of course can mean something different in classical and operant conditioning. In classical conditioning, reinforcement is the presentation of the unconditioned stimulus which elicits a specific and conditioned response. In other words, to remind you of Papa Pavlov's experiment with the bells and the salivating dogs, the sound of the bell was the unconditioned stimulus, which at first caused the dog to form no saliva. But when the experiment had been repeated a number of times, the dog was emitting saliva at the sound of the bell, so the stimulus, i.e. the bell, followed by the reinforcement of the food, caused the conditioned response, the production of the saliva. The sound of the bell had become associated with the presentation of the food.

Pavlov's work is extremely important in training horses, since we are asking them to do things that are completely alien to

71

them in response to alien stimuli. For example, a horse in the wild never jumps, but no matter which method you use to teach the horse to jump originally – what stimulus you use – you can condition him to clear three foot or three foot six without any difficulty.

The old-fashioned method of doing this was to put the horse into a jumping lane and drive him up the lane over low obstacles with a lunging whip. Sometimes the horse would be loose and sometimes on long reins, but whichever method was used the horse learnt to jump to escape the pain of the lunging whip. That is, negative reinforcement was used.

But once the horse was trained, he would continue to jump whether the fear of the lunging whip was present or not.

We, as you know, use the method of getting the horse to jump following other horses, and then we praise him after he has jumped. Here also, once he has learned to jump the fences, the horse will jump whether or not the incentive of the other horses is present. So, the conditioned response, that is jumping the fences, remains even though the stimulus is no longer there. Your conditioning treatment will have made jumping a reward in itself.

This means that it in turn can be used to reward other achievements. You can precede your jumping with a teaching session, for example to do a flying change of legs at the canter, and use the jumping as a reward for new learning. Then, if his jumping entails changing legs to jump a series of obstacles set at different angles, he will soon see the point of doing tight figures of eight so that this lesson, in turn, becomes the reward for learning to go from a walk straight into a canter and straight back to a walk. This again will improve his perform-ance at jumping, since it will enable him to jump difficult fences set at right angles to each other.

From this it can be seen that if the process of learning is reversed so that the ultimate objective (which in this case hap-pens to be the most enjoyable) is taught first in its simplest form, the more tedious forms of schooling can be eliminated. So the horse learns more quickly and, most important of all, enjoys his learning, and instead of dragging hours spent in a dull classroom, life becomes interesting and exciting.

If the horse's learning, or more accurately your teaching of him, is planned in such a way that each new piece of learning can be seen to increase the pleasure of previous learning, or to make a piece of learning easier to perform, you are increasing the horse's desire to learn. This in turn increases his state of arousal so that he learns each new schooling movement faster, and with greater enthusiasm and ease. A horse can see that from one initial piece of learning the whole horizon of his enjoyment is broadened. And in addition each random task of his operant learning, instead of being a matter of chance, is placed in such a situation that it becomes the obvious thing for him to do.

Let me remind you of what operant, as distinct from classical, learning is. The horse is normally being trained in a situation where a number of alternatives present themselves, only one of which is the correct one. When the horse chooses the correct response, he is rewarded, and when he does the wrong thing he is punished; or, alternatively, he is not rewarded. You can, if you want to avoid punishment, present and re-present the horse with the alternatives, and offer the reward only when the correct response is given.

This, clearly, is a slow business. So if the next thing to learn can be seen as obvious following the previous piece of learning, learning can be much quicker and more effective. The example of sequence learning described above illustrates this perfectly. The horse learns to jump by following other horses. This gives him pleasure, so he jumps by himself. In learning to do a flying change of legs he sees that turning corners to approach another fence is made easier. During his flying change schooling, he can see if he drops from a canter to a walk at the point of changing legs, changing legs becomes easier, and putting himself right when approaching a fence becomes a matter of course. So, his jumping in turn again becomes easier and more enjoyable. If each piece of learning is devised in a similar manner to suit each horse and each phase of training, the horse's learning ability increases and teaching him becomes a far easier matter.

I first used what might be termed the motivatory method of
teaching a horse, with a four-year-old thoroughbred stallion we
called George. He was superbly bred, but unfortunately his
mother had died at his birth, so he'd never grown properly, and
by the time we got him he was only 15 hands, which was not
big enough to race and not even really big enough to make
anything better than a second quality crossing stallion. So we
bought him very cheaply and had him cut. My father wanted
him to play polo on. The pony I was playing polo on had gone
lame at the time, so I started to work on him. Then there was a
panic because one of the boss's ponies went lame as well, so
we were very short of polo ponies. This was really the chance
I had been looking for. For some time, I had been working on
a method of training horses using communication, psychology
and learning by behaviourist principles, and I wanted an excuse
to try what was nothing less than a revolutionary method of
teaching horses. In this case I proposed to break and school a
horse to the point where he could play a couple of chukkas of
polo, all within three weeks of his being castrated. So I took
him home, and got straight on his back (this was about a week
after he had been cut), though for only half an hour. To start
with, my wife led me on foot, then she let me go with the horse
just following her, walking by her side. The second day, she led
for ten minutes, then we went back to the yard and she
mounted a second horse, while I rode George.

It was a lovely day and we had a very pleasant ride for about
half an hour, walking most of the time, interspersed with a few
strides of jogging. On the third and fourth days we increased
the jogging to trotting, and on the fifth, when he was walking
and trotting and responding to the bit and reins a little bit, on
our way home we came to a ten-acre field, my wife's horse broke
into a canter, and George followed him. Her horse started
bucking a bit from pure joi de vivre; George bucked in imitation, not trying to get me off but just because he was enjoying
himself. After this we were walking, trotting and cantering each
day, and after about a week, when my wife couldn't come out,

I took George out by myself, along approximately the same ride. As we came near the first of the places where he was used to cantering, he started dancing about and reaching for the bit, so that I had a job to get the gate open and shut. I got the gate only half shut when he was off, really stretching out and enjoying himself, not in the delicate canter that we had been doing, but flat out. As we came towards the end of the field, I wondered what the devil I was going to do, since he was leading with his near foreleg. As we came to the hedge, I threw my weight to the right so that he turned up the hedge. The combination of my weight suddenly coming on to his right side and having to turn sharply led him in to changing legs as we went up the side of the field. I turned him back down the field the way we had come and he went back at a much steadier pace. Halfway across the field I turned him to the left, again throwing my weight and knee sharply over to the left. He changed back from his off fore to his near fore, and for ten minutes we zigzagged about the field. By the time I'd finished I had only to throw myself over and touch the reins and he was turning and changing legs just like that.

I had had great hopes for this method of training, but the spectacular results that I was getting were more than I had hoped for. I repeated the schooling the following day and to my amazement discovered that he was enjoying changing legs on command. At the end of the schooling session his behaviour was like that of a teacher's pet at school who has given all the right answers at exactly the right time: he swaggered down to the yard showing all and sundry that not only was he a very pretty horse, but he was also a very clever horse indeed. And indeed I had made a great point of telling him how clever he was.

Because he was rather tired I felt that this would be a good time to introduce him to the polo stick, which my wife brought out to me. He looked at it with very great disfavour, but I put it round my wrist and just let it hang down at my side, and rode around the field two or three times with it there. Then I started swinging it backwards and forwards very gently and slowly. Each time I swung it he shied violently to the left, but after a bit he got used to it and we just walked round hitting the tops off thistles and buttercups, and he decided that the stick wasn't going to kill him. He relaxed, and let me go on for about five or ten minutes knocking the tops off thistles on the off side, which is the normal side you carry the polo stick, before I lifted the stick over to the other side, doing near side forward and backhand strokes. This time it took him only about five minutes to decide that everything was all right, and this was just a piece of my normal half witted behaviour.

My wife rode with me the next day, and she carried the polo stick to start with, then we changed over. I took the stick, first walking, swinging backwards and forwards, and then trotting, and he took no notice of it. When we got to a place where we could canter I cantered him changing legs, then I came back and got the polo stick and did the same thing with the stick. By this time he was quite used to the polo stick and took no notice of it whatsoever, so when we got home we went on to the next stage of his schooling. We got out an old polo ball, and I tapped it about a bit, on him. At first he looked on it with disfavour, but after about half an hour, he was beginning to follow the ball's direction with his eyes when I hit it, and then to follow it himself for me to hit it forward again. Then I did a backhander and, without being asked, he spun round and trotted back to where the ball had dropped.

The following day we were due to play polo, so I took him down with us, and seeing all the strange horses and cars and trailers was a bit much for him to start with. But I went into the normal schooling routine for ten minutes or a quarter of an hour, tapping the ball about. Then I let him watch the first chukka of polo, which was a fast chukka. The sight of the galloping horses and their keenness and enjoyment fascinated him, and above all gave him some idea of the reason for his learning. I put him into the second chukka, which was a slow chukka, and he took to it like a duck to water, just loving galloping with the other ponies chasing the ball, and leaning against the other horses to ride them off the ball. It was something so new, interesting and exciting that he thought it was absolutely wonderful.

He enjoyed it so much that I let him play in a second chukka at the end of the day, by which time he was becoming a dab hand and was changing legs, twisting, turning and following the ball with the best of them. He found great pleasure in riding other ponies off the ball, so much so that he hit one much respected and ageing Colonel such an almighty clout with his shoulder that I was told in no uncertain terms the Colonel's opinion of me, my heritage and the horse I was riding.

Unfortunately for me, George proved such an apt and willing pupil and within a month had turned into such a good polo pony that my father was offered a price for him too high to refuse, and he was sold and sent up to Cowdray Park, where he played polo for many, many seasons. But this is what happens when you get a good horse – if you're lucky you keep him for a month or two and that's all, because someone else with a much bigger pocket than your own comes along and buys him off you. Whereas if you've got some useless screw who you'd be only too pleased to see the back of, nobody wants him and you're stuck with him for ever.

George's transformation from an unbroken stallion to something capable of playing polo in three weeks, and then into an extremely good polo pony in a couple of months, proved to me beyond all doubt that my theories about using communication and psychology, and making the most of the horse's learning capacity, could make schooling easy and pleasurable for me and my horses. The revolutionary method of breaking I had evolved had worked beyond our wildest dreams.

I had demonstrated first, that by being led and made a fuss of, George had learned to accept me on his back immediately. Second, by following another horse and imitating what his elder was doing, he had learned to turn and respond to the bit, so that when I pulled the left rein and his companion turned to the left he naturally responded to the bit by turning to the left; when I pulled the reins his companion stopped, so he stopped as well. He was being conditioned by association, like

Pavlov's dogs. But he was being reinforced in his learning by being told what a good and clever horse he was, and he was responding to the praise. Third, when I taught him to change legs by making use of the situation I found myself in – going at a rate of knots at a hedge – and altering my balance, I altered his balance so that he did a flying change naturally, this being the only way he could turn in the direction he was compelled towards; and by going then from his off fore, which was his unnatural leg, half-way across the field, turning him and using my weight to bring him back on to the leg with which he led naturally, he learned to do a flying change without any difficulty whatsoever. At this point I was facilitating his operant learning by putting him in a situation to encourage him to make the choice that would be rewarded.

Then, in learning to accept the polo stick, he found that (a) it didn't hurt him and (b) he got praise and reward from me. It gave an added interest to his life when the ball was introduced and he found it quite amusing to follow the white object that was bobbing about the field, and when he got to it, to find that the stick made the ball go bobbing away from him again. He quickly learned to follow the ball. And when he got down to playing polo he found the excitement and enjoyment of galloping, turning and moving with the other horses was such that he increased his own natural handiness and ability without any real teaching. He was learning from his own experience, thinking about the game and enjoying using his brain, his brawn and his speed to excel and beat the other horses. Again, of course, each time he did something well he was praised for it.

One of the amazing things that I discovered in training George, in the two months or so that I had him, was that at no time did I have to touch him with my heels and at no time did I have to hit him with a stick. Occasionally he got a clout across the legs with a polo stick, which happened from time to time when we were playing polo, but he ignored any pain or discomfort, accepting it as part of the game just as a human being accepts the bumps and bangs of the rugger field.

This was our first really successful experiment in allowing the horse to learn instead of teaching it, and it illustrated perfectly how, if each situation is presented to the horse correctly, the horse's learning speed can be increased dramatically.

To achieve the maximum results, the required response, motive and reinforcement must not only follow in correct sequence, but any reinforcement required must be immediate. You can see the sequence in any stable at feed time. When the horse is hungry (motive), you take a feed into the loose box. When he comes to the food (required response), he eats, so satisfying his hunger (reinforcement).

Above all, negative reinforcement must be immediate. To illustrate negative reinforcement paired with response and motive suppose you are dealing with a savage horse who attacks you with his teeth and front feet. The motive here is aggression caused originally by fear. You deal with it by lunging forward, punching him with your fist and roaring at him (negative reinforcement). In every case I have handled in this manner the horse has retreated, usually to the corner of the loose box (required response). In this case the negative reinforcement has diminished his motive to attack. A series of lessons of this kind will soon eliminate the motive to attack, and on your approach alone he will retire to the corner of his box. And provided that at the same time you are feeding him, watering him and spending half an hour a day in the box just chatting to him, you will arouse no further fear.

Negative reinforcement can of course be used to decrease the motive itself. A simple example often arises in shoeing a horse. One of the most annoying habits a horse can develop is to stand up on his hind legs, or hang back, when his front feet are being shoed. Here the motive is twofold, first to annoy you and second to avoid being shod in front. The response required is for the horse to stand still. The reinforcement used is to press the base of the dock against the horse's anus. If he is standing still, the pressure is light and causes little or no discomfort. If, on the other hand, the horse attempts to come backwards or rear, the pressure is increased to the maximum, causing great discomfort.

This makes the horse go forward again and stand in the correct position, providing the motive to stand still and give the correct response.

The third way negative reinforcement is used can be illustrated with a horse out hunting. If a horse looks like refusing a fence (motive, not to jump), you catch him once with your hunting crop (negative reinforcement), so he jumps the fence (correct response) and so gallops after hounds with other horses (here positive reinforcement is added to negative reinforcement).

I remember, a horse called Croix de la Lorraine, who was a useless flat race horse of considerable ability. His owner had purchased him for a fair price (since he'd twice been placed over two miles) to go hurdling. He proved a complete failure, adding to his refusals to gallop or race his insistence that he would not jump. I ended up being given a half share of him, in return for feeding him and curing him of his vices.

The first thing to do was to give him a motive to win. This I did, by galloping him with a series of slow horses throughout his training period, so that he always won. Second, I hunted him, making sure that he never jumped a fence. As soon as he got excited and was enjoying himself hunting (that is, I had induced a state of arousal in him), I would take him home. By this stage I had induced (a) a motive to gallop and beat the other horses, and (b) a state of arousal whenever he went hunting. By now he was tearing at his bit to get in front of other horses, and resisting stopping and going home when I wanted to. So I used the motive to beat other horses and his state of arousal (excitement) to jump four or five fences on my way home. The following day I was hunting, I jumped twelve or fifteen fences before going home. That was on the Monday. Even though he was still refusing to jump anything at home, I ran him in the maiden race at Sparkford Vale on the Saturday, running in a suicidally large field of 34. I led from the start to within two fences of home, where he was passed by the eventual winner. He won easily the following Saturday.

Unfortunately in his fourth race he fell and broke a leg. But in spite of this tragedy, he showed clearly how the introduction of a motive, by using the reinforcement of enjoyment and praise, together with arousal, can not only induce learning but will also lead to the required response.

Motive reinforcement and response is the key to everything that a horse does, even though the response is not always the one required or expected. Salline for instance whickers at the sound of the bucket at feeding time, welcoming her food. But she also kicks her partition. Here you have two responses to the same stimulus, the first to welcome you, the second (kicking her partition) to tell the adjacent horses to get out of the bloody way since she is eating. This is an example also of pairing a stimulus (feeding) with a response (kicking to stop other horses eating her food), even though she has been fed by herself for six months and no other horse has had the opportunity of stealing her feed.

Another example of this pairing of response is the hunter who will become excited at the sound of the hunting horn, because he associates it with hunting and pleasurable excitement, even though he has long ceased hunting. In the case of one horse I had, paired response nearly ended a racing career. Molfre loved racing, and he used to get very excited when we got to the track. One of the symptoms of his excitement was that he would break out in a sweat. He very quickly learned the preparations for racing day – the day before he was groomed and spruced up, his mane was got ready for plaiting and his tail was bandaged, and at ten o'clock at night his hay net and water were taken out of his stall.

Then as soon as he had been put into the lorry he would start to get hotter and hotter, and before we had gone twenty miles he would be in a muck sweat. Then things got worse, because the night before racing, after we had got him ready for the following day, he would spend the whole night walking round his box so that he was almost exhausted before we left for the race meeting, let alone before we started the race.

In fact, within the first four or five meetings it became obvious that the only way we could get him to a race meeting fit to run was by getting him ready not the night before, but at the last possible moment in the morning. We also took to driving him around the countryside every now and then in the Landrover and trailer, spruced up as if he were going racing, so that after a series of such excursions which ended up at home and not at the racetrack, he ceased to pair racing preparation with racing, and he never knew whether he was going racing or not until he actually got to the meeting.

In this case a horse was associating the preparations for racing with the racing itself, and exhausting himself with his own excitement. Another kind of pairing can arise when you are trying to get a lazy horse to jump. You may have to tan his backside to start with, but he will quickly learn that he has to jump and he will jump to avoid being belted, so that you do not have to touch him at all when you want him to jump: he has associated going into a fence with avoiding a bleeding bottom by jumping. Here the negative reinforcement is paired with approaching a fence. This in turn will be reinforced by the fact that he actually enjoys jumping, especially if he is jumping with other horses.

I carried this tendency to an almost comic extreme when teaching Dart to jump barbed wire. Because he approached his fences at speed he wasn't always seeing the wire, so I got into the habit of kicking him and saying 'hup', when I was approaching a wire fence. This finally reached the stage where I could gallop across a field, kick him and say 'hup', and he would jump three-foot-six, even though there was no obstacle or sign of obstacle present. I did this quite simply, first of all by putting a rail between two posts with a thin wire rope three inches above the rail, kicking and saying 'hup' on his take off stride. I then removed the rail and jumped him over the almost invisible strand of wire rope, which was by then set at three foot nine inches. I then proceeded to move the wire rope round every fence so that it was anything up to two feet above the rail which I had on the fence already. When he was jumping the wire rope on command, I started stretching it across the field between available trees and posts, which could be anything up to thirty feet apart. Eventually, I discovered the wire itself was unnecessary because he was jumping on command automatically.

A stimulus that has become associated with a primary reinforcement in this way, is called a secondary reinforcement.

Pairing the reinforcement with the response to be learned is an essential requirement for both classical and operant conditioning. Many studies have indicated that the best interval for such pairing is about half a second. Long delays between the response to be learned and the reinforcement greatly retard the learning. So the main problem is how to get the correct response and the reinforcement closely paired, and one solution to this problem is to combine a secondary reinforcement with what is called 'shaping'. Shaping is rather like the game of hunt the thimble, in which someone in a room is trying to find a hidden thimble and the spectators, who know where it is, say 'warmer' when he gets closer and 'colder' when he moves away from it. In such a case the spectators are immediately reinforcing any response in the right direction, thus strengthening that response. Shaping keeps the subject closer to his goal rather than wandering around until he accidentally finds it.

Shaping initiated by secondary reinforcement promotes very rapid learning. A simple example of this is teaching a horse to be caught. First you put food in a bucket and leave it in his field: very quickly he will learn that the appearance of the bucket in his field means that food is coming, and he will come towards you as soon as he sees the bucket. After you put the bucket down he will come to it and eat from it. But in the second stage of your shaping the reinforcement, i.e. the reward of food for coming to you, he has to eat out of the bucket in your hand. In this you are shaping him in the direction you want to go, i.e. to catch him. In the third stage he gets his reward of food only when you have your hand on him, and in the fourth stage he gets his food only when you actually catch him.

You can see how you have shaped your reinforcement. To start with he gets food for eating out of the bucket, then he gets food for allowing you to put your hand on him, and finally you have reached the desired goal which is to give food only when he is caught.

When shaping the responses of Charlie, the flat race horse I bought at Ascot, I was aiming to get him to jump steeplechase fences. When we bought him he had never even walked on rough ground, and to get him to cross a slight depression was a major task. But within a day or two he was quite happy going over slight depressions, and each time he stepped over a dip in the ground he was made a great fuss of. The next stage in shaping him to jump was to run over a very small ditch, about a foot across and six inches deep. This was quite simple for him to jump, but it took at least twenty minutes before I could get him to take his courage in both hands and step over it. After he had done this I made a terrific fuss of him and after another five minutes' coaxing he stepped back over it and back and forward until he was just stepping over it without thinking about it. Then I told him how clever he was and took him home.

The following day I got him stepping over a bigger ditch, and by the time he had finished on his second day of learning to jump he was actually jumping a ditch about two foot across and three or four feet deep. The bigger the obstacle the more he was told how clever he was. On the third day, I looked for all the little ditches I could find on the mountain and he was really enjoying popping over them.

The next stage was to get him to take off from the ground over a solid obstacle. Normally, of course, I would have had him hunting or going up a jumping lane behind other horses over a low obstacle. But with Charlie this wasn't possible, since he had been racing and was a sprinter, and when he was with other horses he would tend to try to race them. This meant that he would be thinking about racing horses and not about jumping fences, and there was a very great risk that he might fall over something and hurt himself, which would have put him off jumping for life. So, he had to learn to jump by himself.

On the fourth day I took him into the forest where I knew there was a small tree trunk about a foot high across one of the rides. I trotted him down the ride until we came to it. When he saw it he stopped as if it was the most frightful object he had ever seen in his life, but after about ten minutes' coaxing I managed to get him up to it. It took me another quarter of an hour to get him to step over it, and then again he was walked backwards and forwards over it until he was thinking nothing of it. I went on about another half mile to where there was a tree about eighteen inches across, lying on the ground. This time he had actually to take off, and this he did. Again I took him over it several times, and back over it again. He suddenly found that this was fun, and the fifth or sixth time when I turned him to pop over it again, he spun round on a sixpence and flew over. I told him how clever he was and that was enough for one day.

By the end of the fifth day he was jumping about two feet and really enjoying it. On the sixth day a friend took him out hunting and then Charlie began to understand what life was all about.

This is how you shape a horse – you encourage him. We started off making a fuss of him when he first stepped over a ditch a foot wide and six inches deep, but within six days he was finding that the reward for jumping was the jumping itself: we had shaped his desires so that he really wanted to jump and he was now loving his jumping and thinking it great fun.

The important thing when training a horse is that the primary end should be that the horse is enjoying doing what you want him to do. I realize of course that in every case a horse has to do a large amount of tedious work, but it must always end with the horse enjoying himself.

9: *Retraining Techniques: Weakening Learned Responses*

In training your horse you are not only looking for enthusiastic responses to what is to be learned, you also have to weaken the undesirable responses which have been previously learned. This is an important matter because responses that have been learned over a period in everyday life take a considerable period of time to weaken. Many of the habits that your horse has acquired are not desirable, and you will want first to weaken them and then to get rid of them altogether.

How can a learned response be weakened? One way is simply to allow the horse to forget it, by not allowing him to get into the situation that will provoke it. If, for example, a horse has got into the habit of bucking when you hit him, you can stop him by not hitting him. When he has ceased to associate being hit with bucking you can start using the stick on him again.

The second way to weaken the response is to extinguish it. Extinction, the key term here, is both the procedure used and the result of that procedure. The critical feature is that we stop reinforcement of the behaviour that we wish to extinguish, by punishing the response. For example, with dogs you can extinguish the behaviour pattern we described earlier as conditioned in response to the bells, by ceasing to reward the dog with food when the bell rings. If your dog salivates at the sound of the bell, you can ring the bell continually without feeding him, and he will cease to secrete saliva. Similarly with the horse who bucks when you hit him, to get him to forget about bucking, you can hit him every time he bucks when you take him out. You will extinguish the desire to buck simply by changing his association of ideas: whereas previously he associated bucking with your stopping hitting him, he will very quickly discover that his

86

bucking makes you hit him and so he will cease to buck. With yet a third method of curing bucking, it is advisable to point the horse up a steep hill. This has two advantages; first this way it is much easier to stay on top, and second he will tire of going into orbit much more quickly. If, on the other hand, you point him down hill, even if you do not pass a rocket on its way to Mars, both you and the horse are likely to go base over apex at the bottom of the hill. I had a very vivid example of this a couple of years ago when Clancy, after jumping a large post and rail, put in a small buck of joie de vivre which took us both over the edge of a mini Mount Etna. When we hit the bottom we both of us bounced back up into the air, turned three somersaults and landed up to our bellies in freezing water.

The method of extinction we use with a bucking horse is to assume that, other than the buck of joie de vivre, a horse bucks to annoy you. To cure his bucking, we take him out riding to the bottom of a steep hill and we point his head up the hill. I drive my heels into his side and say 'right, go on buck, you beggar', and of course having two heels rudely, painfully and for no apparent reason belting his ribs makes him buck. Each time he lands from bucking, I drive my heels into his side again and repeat the process. He will do this four or five times, but of course by the time he has bucked half way up the hill he will find the effort of bucking quite considerable and since I'm telling him to buck, his natural bloody-mindedness makes him say 'right, I won't buck'. Two or three days of this treatment extinguishes all desire to buck. This method has the advantage that (a) we do it in a forest drive, which is comparatively narrow, so that he has to go straight up the hill, and (b) it's much easier to sit a buck going up hill than on the flat or downhill.

We use the same method to stop a horse kicking. When, for example, we get a horse that kicks when you touch its legs, I get in close to the horse's side, run my hand down his leg until he kicks. As he brings his leg forward towards me I push the leg away from me, so that when he kicks back he is quite likely to kick his other hock. Then I say 'good boy, kick—go on, kick again' and run my hand down his leg. I go on doing this until he is tired of kicking himself. But if, on the other hand, I was afraid of his kicking, and didn't want him to kick, the very fact of my obvious nervousness would tell the horse that I was frightened of him, which would give him an incentive to frighten me more, by kicking again : and since in my fear I wouldn't get tight into his side, I would probably end up in the ideal position for him to bash me through the stable wall. By using the horse's natural bloody-mindedness, which all horses with a vice have, and by telling him to do the thing that until then no one has ever wanted him to do, you extinguish his desire and his motive to kick, or to buck.

Where fear conditioning has been used, that is a horse has learned something because he has been frightened, it is much more difficult to extinguish the fear and the association, than it is where the learning has been by positive conditioning, i.e. where a horse has been rewarded for learning something. A horse that has been rewarded for learning will forget about it in approximately the same length of time that it has taken him to learn it, but where fear has been used it will take considerably longer to extinguish the fear than it did to instil it in the first place.

A typically stubborn example of fear conditioning was a horse called Kingsmoor. I'd seen him first in the autumn when we were hunting between Langport and Mutchenly. He stood out from the half dozen horses he was with, a big, rawboned, slashing bay horse of terrific quality, and I enquired about him from my next door neighbour. It seemed that he was an eight-year-old, and that he was unbreakable, untouchable, unmanageable. Anyone who was fool enough to try to catch him must be a fugitive from a lunatic asylum, and, if they did get to handle him, wouldn't live long anyway.

This was just my cup of tea, so I made sure that I bumped into the owner of the horse accidentally two or three weeks later, talked to him about this and that, and mentioned that I'd seen the horse and liked him. He quickly ordered me a series of double whiskies one after the other, extolling the virtues of the horse but failing to mention any of the vices that I already knew about. In the end, I bought him for a tenner, which was below his killing price at the time, the only condition being that I should catch him myself.

It wasn't until the middle of February that I could make arrangements to fetch him. I telephoned his owner, only to be told that I couldn't possibly get him because he was down on the far end of the moor and there was a mile and a half of water between him and the nearest road. The remainder of February was extremely wet, so it was the second week in March before it was anywhere near possible to fetch him. I hired a lorry and off we set to catch the horse. We got to the farm, paid for him, picked up the owner and drove a mile and a half to the nearest point to the horses we could reach in the truck.

The willow trees were just beginning to burst into bud, with their delicate greeny-yellow leaves like long thin hands in pale green gloves. There were great lakes of water, left from the flooding, shimmering silver in the sunlight. Here and there ash saplings were standing slim and slender to attention, as if saluting the sun.

The only way we could get to the horse was to walk along a ridge of high ground, and then to wade two hundred yards down a lane flooded well up above our wellingtons. We got about a quarter of the way when the lorry driver said he was damned if he was going any further, and the girlfriend I'd brought with me was wilting considerably, but since at that time we were very much in love she would put up with anything. Three parts of the way there, the owner suggested we turn back and leave it for another day. But as we were nearly there, we kept on. When we finally got to the field I had a nasty shock: the horse and five ponies had been marooned for a month on about half an acre of ground, and the spectacle that greeted us was of six living skeletons looking hardly able to move. We opened the gate on to the lane and proceeded to shepherd them towards it. We got them half way down the field when they broke back, galloping along the edge of the water, so back we squelched with our boots full of water and drove them down again. We did this three or four times before their general weakness overcame them and they allowed us to drive them on to the lane.

Here we were lucky because the big horse had naturally established his leadership, so when we came to the spot where the water was a couple of feet deep in the road, he just splashed through it. The five ponies decided it was too much of a good thing and refused to go any further. So without much difficulty I edged round them and, leaving the others to stop the big horse coming back, I drove the five ponies that we didn't want back into their field. We then drove our horse on down the road to where the lorry driver had backed the lorry into a narrow piece of the lane and arranged it very skilfully so that the horse couldn't get past. In about ten minutes we had walked him up into the lorry without frightening him, put up the tail board and set out for home.

When we got home I changed into dry clothes and my girl-friend borrowed some of my sister's. She was only five foot two and my sister was about five foot eleven, but by tucking bits in and rolling bits up she was able to make herself reasonably mobile, so we went to unload the horse. I opened the door at the front and got myself halfway in, to be met by a tornado of flying front feet and teeth. I came out through the trap door like a shot from a gun, counting myself lucky to have sustained nothing worse than a bleeding hand.

This was a horse who went for pure attack – unprovoked, and at the very sight of a man at close quarters. Clearly a certain amount of thought had to go into the next move. I had first to halter him, then to get him out of the lorry into the stable. So we backed the lorry halfway up the yard towards the stable, and dropped the tail board. Kingsmoor – that was what I had christened him – stood up in the corner of the lorry while we opened the gate. When we got to the second gate he made a rush for it. Quickly the lorry driver and I shut it hard. Then we started to open it again, slowly. Again he attacked us. This happened four or five times before I was able to open the gate enough for me to get to one of the partitions that divided the lorry. I got it unclipped and gave it a push in towards him. I did the same with the top part of the partition, and went in very very slowly, pushing the partition away from me all the time. When Kingsmoor next made an attack, I simply flung the partition back a bit so that I was protected by it, and put my hand round it to touch him. Since he was unable to get at me he was left with no choice but to let me handle him, or to go right backwards into the other half of the lorry. This he did, so I pushed the gate shut again.

At this point my father returned, so we had our usual blazing row about how incompetent I was, but since this was the first horse I had ever bought he eventually consented to allow me to make a fool of myself in my own way. Together he and I eventually got the two partition gates shut, which meant that we were in half the lorry and Kingsmoor was in the other half. By edging the gates forward until Kingsmoor was tight up against the top end of the lorry, and at considerable danger to ourselves, we got him eventually into the position I wanted him in, and I fetched a halter, a box and two plough reins. I stood on the box and leaned over the narrow space between the top of the lorry and the partition. Kingsmoor tried to bite me, but he couldn't get his head far enough round. I tried to ease the halter over the far side of his head, but he was having none of that and I couldn't reach, so I shifted the box and tried again. Kingsmoor could reach me here: he made a dive at me and I jammed the nose-band of the halter over his head, at the same time quickly putting my hand over and catching hold of the strap which I buckled up. His teeth had bruised my arm, but I suffered no other ill effects. The next thing was to tie the two plough reins together. I fed one end of them through the side of the gate and tied it to the cheek piece of the halter. The rest was easy. The two plough reins together made about thirty feet, and the far end was given a double twist round a post.

Now we were ready to open the tail board and the gate. Kingsmoor made one leap, jumping from the floor of the lorry into the yard, and went straight up the yard as if his life depended on it. Unfortunately for him, he was travelling at a fair pace when he came to the end of the halter rope, which jerked him round and his feet shot from under him. My father, who was closer than I was, made a dash and was sitting on his head before he had a chance to get up. We undid the plough rope from the post and ran it through the ring in the stable where we planned to put him, and let him get to his feet. Playing him rather like a salmon on the end of a line, we slowly got him nearer and nearer to the stable as he dashed from one end of the rope to the other. When at last we got him into the stable, we put up two slip rails so that he couldn't hang back too far or get out, and left him with a very large feed and a large amount of hay and we all went for tea, wondering what the hell we were going to do in the morning when it came to feeding him again.

What we did when we tied him up in the evening, was to run his rope through the ring on the manger, back to the side of the stable. The following morning, when it came to feeding, I untied the rope and eased it back so that he would swing round. Immediately, he made a lunge for me with his teeth. I jammed the feed bucket over his nose so that he had a mouthful of his breakfast instead of a mouthful of me. I kept the bucket there so that he couldn't bite me, and he quickly got the idea and started eating. I took it up to the manger and tied him up very short again, so that he couldn't bite me, and then slowly and quietly, while he was eating his breakfast I stroked and teased the worst of the tangles out of his coat.

After a while he relaxed a bit, and by the time an hour had passed had been convinced that fear and anger were not the only appropriate reactions to human beings.

During the night I had been thinking about Kingsmoor's aggression, and I had concluded that he was attacking people not simply out of anger, but because anger was the best form of defence against fear, and I thought that if I could eliminate the original fear I would also eliminate the anger.

This proved to be the case. When I went to him after having my breakfast, he was already less afraid of me, and although he made a couple of kicks in my direction, he didn't attack me at all. From that morning on, other than snapping at me when I was grooming his tummy or doing something that irritated him, he never attacked with his teeth and front feet again.

I took an old saddle with me which I eased on to his back and girthed up by degrees, little by little over half an hour, until it was firmly placed. I would tighten the girth one hole, make a fuss of him, and so on until I'd got the saddle on properly. Then, using the leather halter as part of the bridle, I got an old vulcanite snaffle we had, and tied it to the cheek pieces of the halter, putting the bit in his mouth. With a halter rope which I made into a rein, I now had the horse ready to ride. My father got back for dinner at about half past twelve, and I asked him to give me a hand up on to the horse, which I did in the stable, both the boss and myself keeping well out of range of his hind legs. I'd put a feed in front of him, and funnily enough when I got on him, he barely took any notice at all, so I got off and on several times, and he stood like an old horse.

We went and had lunch and when we came back I got on to Kingsmoor again and my father untied him, taking the two plough lines and knotting them round a post, then slipping one through the bit to pull Kingsmoor round, very very slowly. He came round and took a dive through the stable door, taking the skin off my knee, nearly decapitating me and tearing my coat on the top of the stable door as he went. The idea had been that my father should lead the horse, but the sudden tug of twelve hundredweight on the ploughline caused him to let go, and I was on my own.

Kingsmoor made a dash down the yard, completely ignoring my efforts to persuade him that it might be a good idea to steady down, out through the front gate and down the road towards the river. But after about five hundred yards his weakness caused him to slow a little, and before we reached the river, half a mile away, he was back to a trot, and several hundred yards further on he steadied to a walk. I let him walk on slowly, picking at the grass as he went, making a fuss of him and talking to him. After about a mile I turned round and we meandered home slowly, meeting the boss about halfway home. He had followed me in the car – evidently calculating that the car would be handy to pick up my body and take it straight to the morgue.

We got home and I rode Kingsmoor to the stable and slipped off. My father had thoughtfully put a good feed in for him, and without any difficulty at all I tied him up, took the saddle off and proceeded to rub the worst of the sweat off his soaking coat. As I moved towards his hind legs he lashed out at me, but except that he hurt himself a little when his leg hit the wall, no damage was done. Being a clever and thinking horse, he made only two tentative kicks, hitting the wall each time and not touching me, so he settled down to the serious business of eating his dinner.

I fed him at six and again at ten. The following morning, when I went in with his feed he whickered with welcome. I was under no illusion that he was welcoming me – he was merely greeting his breakfast. But over the next six or eight weeks, he gradually stopped kicking altogether. After the first week my father was feeding and handling him as well as me, and soon my sister was handling him too. Meanwhile of course he was in, and being fed well, so as he got bigger and stronger he had a few tries at bucking me off. But he wasn't too serious about it, and it was clear that once the fear syndrome had been extinguished his anger at being approached by human beings had disappeared with it, and so had his desire to misbehave. The technique we had used to extinguish the fear was to substitute, i.e. to replace it with another feeling, that of pleasure. We had made him pleased to see a human being, by making use of his desire for food. Since the appearance of the human being was always accompanied by food, he was pairing the pleasurable anticipation of eating with my appearance. This association was closely followed by others – the pairing of my appearance with the excitement of exercising, jumping, galloping and seeing new things – so that in two months he was whickering when I went into the stable whether I had food with me or not.

We had extinguished the fear and had substituted pleasure. But it is always very difficult to eliminate fear altogether, and Kingsmoor was always just that little bit frightened of strangers, and even when he left us about four months later, if a stranger went into his box, he would still hunch himself up into a corner and turn his backside towards them. On the other hand if they stood for a minute or two and talked to him quietly, he would soon relax and lengthen himself and turn his head towards them.

He went on from us to make an outstanding hunter. The man who bought him from us kept him until he died at the age of twenty-two.

It was only after he had left us that we learned his earlier history. He had been born in 1941 out of a hunter mare by a thoroughbred stallion. In 1945 someone had tried to break him, got bucked off and given him a thrashing, got kicked and given him another thrashing for that. After a week of that treatment he had gone back to his owner as unbreakable and had then been sent to someone else. By the time he got to us four people had attempted to break him – one had ended up with a broken leg – and Kingsmoor had firmly associated man with ill-treatment and pain. Yet once that fear of man had been eliminated, and once I'd managed to associate his work with pleasure, because I got him hunting and jumping as quickly as possible, his behaviour pattern changed completely. When the man I sold him to discovered his history, he found it hard to believe, and insisted for a very long time that his informant had got his horse confused with another one.

So Kingsmoor's case illustrates very dramatically both how easy it is, with ignorance, to instil fear into a horse and how difficult it is to eradicate it. Fear probably takes three or four times as long to extinguish as does any pleasurable form of learning.

Where avoidance conditioning has been established – by

which we mean a horse has learned a way of successfully avoiding a task (he avoids jumping a fence by refusing or running out) – this too is very resistant to memory extinction. In some animals it may not be extinguished after a thousand trials, if fear conditioning has been used to establish the avoidance. This was found in those experiments with dogs already described where they were required to jump over fences when a buzzer sounded. As you may remember, the dog was trained to jump when a buzzer sounded, followed by an electric shock. The dogs continued to jump the barrier even when no electric shock was administered in over a thousand trials. It turned out that the only way to decondition the dog, i.e. make him stop jumping the fence at the sound of the buzzer, was by fatigue : i.e. only when the experiment was carried on for such a time that the dog became too physically tired to jump the fence at all, did he discover that the sounding of the buzzer was not followed by an electric shock. The following day he showed no response to the buzzer.

This technique is particularly relevant to dealing with horses that shy. If you have a horse that is in the habit of shying at certain objects or at certain places, you cure him, with a great deal of patience, by coaxing him gradually to approach the object. Or you can ride him past the object when he is tired; and the horse discovers that there is after all nothing to fear.

There are horses however who shy not out of fear, but simply in order to annoy, and in this case the best cure is to erect a large number of alarming objects (plastic bags, rattling tins, etc.), then to ride or lead the horse backwards and forwards past them, using punishment as negative motivation, so that he ceases to shy because he is afraid to do so. But I must emphasize that this method is only to be used where the horse clearly uses an avoidance technique as an irritant or out of bloody-mindedness.

Use of fatigue to weaken an established conditioning is the basis of a number of methods of breaking unbreakable horses. In this country for instance the lunging method is used. In this

Learning to Jump

Two horses and their riders, happy and excited, approach the jumping fences.

Photographs by Terry Williams

The young horse follows the experienced one over the jumps – low, unfrightening ones at first.
Soon the young horse is ready to jump on his own – he has discovered that jumping is fun.

Conditioning Theory

In this experiment, the horse has been conditioned to associate the sound of the blowing of a hunting horn with the appearance of food.
Immediately the horn is blown, he pops his head expectantly out of the stable door.

And is rewarded with his dinner.

Overcoming Fear . . . and catching your horse

A horse registering fear and alarm at the appearance of a stranger in his field.

She approaches quietly and slowly, with a bucket of horse nuts.

The horse approaches . . .

And allows himself to be caught and haltered.

Bad Habits: 1. Napping

This horse expresses his reluctance to cross a ditch, but his rider is firm, and he steps into it.

Positive Reinforcement

He is rewarded with generous praise and patting, and concludes that ditches are not unpleasant after all.

Negative reinforcement (above left)

– whip to be used sparingly, if at all.

Bad Habits: 2. Avoidance technique – running out:

Above right: The horse is able to 'run out' because the rider is going into the jump on a slack rein.
Below: The same jump, this time presented correctly, the horse held together with hands and heels.

Eager horse

Rostellan is groomed, saddled and ready to go out. But Leslie is keeping him waiting, so he decides to let himself out of his stable, and go and find her.

Hunting morning. Leslie is again keeping him waiting, so he decides to –

. . . load himself in to the lorry.

 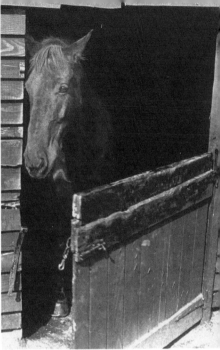

Mischievous horse

Here he is using his skills because he is bored and he thinks no one is looking.

First he lets himself out of his stable.

Then (below, and above right) he decides to open the front door of the house . . .

and see what is going on inside.

Alert horse

Looking forward to a ride.

Bored, dejected and unwanted

At the horse sale.

Choosing Your Horse

Top right: Alone and unhappy.
Centre: Alert and looking
forward to the future.
Bottom right: Frightened
and worried.

(Below and right-hand page) Horses being paraded at a horse sale.
Horse and rider bored with waiting.

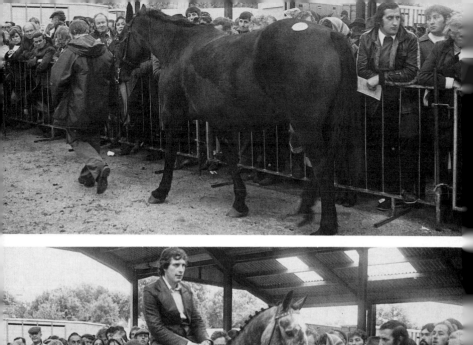

Untouched horses in the herd.

The sheer pleasure of horse and rider alone on a Welsh mountain.

method the horse is deprived of water for three days first, then it is lunged for half an hour to forty minutes. A horse that is 'unridable' because he bucks or rears, for instance, will be so exhausted by this treatment that he will indeed allow the monster who is breaking him, to ride him. But anybody who uses this method is, to me anyway, a monster, and I use the term 'breaking' on purpose, too.

The Irish method of ringing is similar in principle, and cruelty. Here someone takes the horse into a very boggy field and drives it round until it is so tired that it can barely move. Then the man gets on his back and rides him. And the Australians tie the horse up in the sun for twenty-four hours : they too mount it when it is in the last stages of exhaustion.

In each case the horse had been brought to the last stages of exhaustion by fatigue or thirst, and his spirit is then completely broken. Any desire to buck or fight or rear is completely extinguished.

Another way of extinguishing a response is by suppression; this is really passive avoidance learning that weakens a previously conditioned response. Suppose you had a horse that came to you and took food out of your pocket. He would continue to do so as long as there was food in your pocket. You could of course stop him by ceasing to put food in your pocket, so that he eventually gave up looking. But you could also do it by suppressing him, that is by slapping him on the nose every time he went near your pocket. Either way you would be able to cure the habit, but the thing that complicates extinction and suppression of response is the tendency of such response to recover spontaneously. The habit may recur so whilst you may for instance have cured a horse from refusing his fences by suppressing the response, he may start refusing again, if he is weakly ridden, four or even five years later.

Pavlov noticed that a day or so after a series of extinction trials on the dogs conditioned to salivate at the sound of a bell, the salivary responses bounced back even stronger than they had been at the end of the original experiments. He named this phenomenon 'spontaneous recovery'. It is a kind of forgetting in reverse, a tendency to forget that extinction training has occurred. Spontaneous recovery, the slowness of the extinction process and the ineffectiveness of some forms of punishment, all combine to make extinction training a laborious process.

We find spontaneous recovery in horses that buck. After we have taken the horse up what we call our bucking slope four or five times, he will cease to buck, but then after three or four weeks he may start to buck again. Certainly a horse that has once discovered that he can get rid of his rider and avoid work by bucking will, if he goes to a weak rider, immediately start to buck again. Many vices have this tendency to reappear when the horse passes into the hands of a weak rider.

Another factor retarding the elimination of responses is the effect of partial reinforcement. For example, if you are trying to stop a horse taking something out of your pocket, in theory you should repel the horse every time he does so. But in actual fact on some days you will be far more tolerant with the horse than you will be on others.

This inconsistency tends to make the curing of vices and habits much more difficult. Punishment of misbehaviour must be consistent for it to be effective.

It is in fact neither possible nor desirable to reward a horse with praise or titbits *every* time he does the correct thing; or to punish the horse every time he does the wrong thing. Inevitably you will be using some form of partial, or irregular, reinforcement. This can either be random, that is you praise him sometimes when he does the correct thing and punish him sometimes when he does the wrong thing; or it can be what is called interval reinforcement, which means you praise him after every fourth or fifth correct action, perhaps, but at definite intervals rather than random ones. This method is probably more effective than regular reinforcement, at the later stages of training.

Imagine a rider who, riding along, goes kick, kick, kick, kick and then tap, tap, tap with his whip, all the time the horse is walking or trotting. The horse becomes so accustomed to the accompaniment of the drum of the heels and the tap of the whip that he learns to take no notice of either. This regular reinforcement has been completely nullified by constant use. Whereas one sharp cut from the stick or an occasional well driven thump with the heels will have a far greater effect, as the horse will never be absolutely certain when he is going to be punished for being lazy.

Equally, if after each time a horse has jumped a fence you stopped and gave him titbits and made a fuss of him, apart from the fact that you would never win a jumping competition because you would be eliminated for time, your horse would come to expect praise and reward as a matter of course.

What is more, if your horse gets abundant praise and reward each time he does the smallest thing right, you will have nothing more to give him when he makes an outstanding effort. In the same way, if you beat him to within an inch of his life for minor infringements, it will be impossible to give him adequate punishment after a major crime. From this it can be seen that constant praise or punishment soon loses its value; and both partial and graduated reinforcement are essential in training a horse.

A good example of how this works was my training of Royalty, the chestnut mare that I use to hunt hounds. She loves her work : her love of the hounds and pride in being the huntsman's horse are constant reinforcements for her. As the focus of attention at the meet, she swaggers and cavorts as she leads the hounds to the first draw, then she immediately settles down to work, even though she is still a very green hunter. She will stand absolutely still when necessary, and as soon as hounds find, she is watching and listening to see where they are going. Very often she will see the fox before I do. This is all a constant reinforcement for her. On top of that, when she negotiates a particularly large obstacle I pat her neck and tell her she is a good girl, and if she does something wrong or rather stupid I may or may not curse her. Now, this is very mild cursing and the praise too will be mild, so that when she jumps a particularly big obstacle or negotiates a very difficult and boggy place, I can give her extra praise, and she knows she has done really well. And the same principle applies to the punishment, I increase the punishment according to the crime.

10: *Changing Ingrained Habits*

The elimination of bad habits is a major preoccupation among horsemen, simply because most of us are bad riders at times, and thus teaching our horses bad habits. We can, as I described in the last chapter, try to make the horse forget his bad habits. But another way is to replace an old habit with a new one, by making the horse want to do the new thing very much. Replacement with a new habit makes it much easier to eradicate a previous pattern of behaviour. If, for example, a horse's desire to buck can be changed to a desire to jump, and when he jumps he is rewarded, he will cease to want to buck – or that is what you hope anyway.

Suppose you have a horse who has been refusing to jump, you can re-site the jumps so that he is jumping towards home and he has to cross the fences to get there. Then, when you are exercising him, before taking him home you jump the fences and then he goes home to dinner. Not only are you praising him for jumping, you are changing his reluctance to jump to a desire to jump, since jumping the fences will be associated with going home. A horse that has previously refused to jump will often learn to jump very quickly this way. (Of course, here as in any lesson in jumping, the fences must be small.)

If on the other hand the behaviour to be changed is a strongly ingrained habit – ingrained because it has been well rewarded – the habit will be very difficult to eliminate either with punishment or with praise. This was obviously so in the case of Watch, who I bought in Llanybyther in February 1976 with the idea of point to pointing him. I hacked him the first couple of days, just walking and trotting to get to know him. On the third day I decided to give him a canter along the side of the forest. After about four strides he lifted his head sharply, bashing me in the face and practically knocking my teeth down my throat. Using this method he wrenched the reins out *of* my hands and carted me well and truly, only stopping three miles further on, going up a very steep hill. The second time I tried to canter him, he got away from me using the same method, and I only stopped him galloping over the edge of a quarry by changing his balance and turning him base over apex into some trees. He bolted with me regularly for about two months.

This was a typical instance of insufficient negative reinforcement: he disregarded the use of the bit because the reward – galloping flat out, which he enjoyed doing – more than compensated for any pain in his mouth of iron. I cured him in two ways, first by allowing him to gallop on a loose rein at preselected places, and then by galloping him over a fixed timber one-day event course. The first couple of times we did this we went head over heels because we were taking fences too fast, so he learned to steady and place himself at his fences naturally. It was only a short step from here to my pulling the reins and saying 'whoa' as we approached each fence, since the reward for steadying when told to do so was greater enjoyment for Watch. By using this method, he responded at all times.

The treatment, before I had a complete cure, took over a year, but I eventually was able to race him with reasonable success.

Another horse who acquired an annoying habit was my wife's cob Rostellan who, by watching another horse, learned to open his stable door. This was quite an amusing trick; when he learned to open the gate out of the yard and then the gate into the house and garden, it was still quite an amusing trick. So we told him how clever he was in opening the stable door, and when he came to the front door of the house, after opening the stable door, yard gate and front gate, he was given titbits and allowed to wander around and pick at the grass on the lawn. So opening the gates became a well reinforced habit with him. But when he started not only trampling over the lawn but eating the flowers and lettuces, we decided to stop him.

This was extremely difficult, because even if we hunted him

out of the garden whenever we saw him, he was still getting rewarded, because he was getting a certain amount of the sweeter grass around the house anyway, and a certain amount of freedom before we saw him. Even though he was being punished, the habit became extremely difficult to eradicate.

Similarly, when a horse bucks and gets its rider off, it achieves not only its goal but also a certain amount of freedom, which is an additional reward. Every time it bucks its rider off the habit of bucking is reinforced – the horse achieves its goal and gets rewarded. So the habit is very hard to cure indeed. Another problem with curing a habit is that any small rewards and punishment regularly administered come to be taken for granted, so the rewards don't seem as attractive or the punishments so terrible as when they were first given. This is the phenomenon of adaptation to reinforcement, a long phrase which means quite simply that the horse gets used to constant punishment or constant reward, and ceases to take any notice.

A typical example of a horse becoming accustomed to insufficient punishment was Donna. Whilst she was an extremely good pony, being both quiet and of extremely high ability, she became sluggish and bored, mainly because she had been ridden by poor riders. They had used their whips as tickling sticks, going tap, tap, tap with them, and ridden her as if they were rowing a boat with their feet, with their heels constantly digging Donna's ribs. This had to stop. First I changed her feed to racehorse nuts, and increased her concentrates slightly. Second, after a week's increased feeding, I rode her myself for three or four days. The increased feeding had made her a little more lively and as I left the yard, I caught her two with my racing whip, one on either side of her flank.

This immediately changed a tortoise into a torpedo. I steadied her back to a walk in ten or fifteen yards and Donna danced down the road, a thing she had never done before. The laburnum trees along the hedges of our farm were just coming into bloom and when we got to the far side of the valley I turned to look at the breathtaking sight of rows and rows of pale green branches dripping with gold.

Unfortunately, Donna was impatient, so I turned to go on. Donna by now had got the idea: as soon as I slackened the reins she swung up the hill in a spanking trot. At the top, there were posts and rails leading into the forest. I had intended going past them, but Donna, determined to make the most of her new found youth and enjoyment, swung into it before I could stop her and flew over them like a bird. She cantered up the ride for about twenty yards, where we stopped to watch the rest of the party doing their best to follow us.

For the rest of the two-hour ride, Donna seized on every possible obstacle – no matter how difficult or awkward – to alleviate her excitement, and the rest of the riders had no peace. And she frightened the living daylights out of me on one occasion, by flying a tree trunk on the verge of the forestry road, which lay twenty feet below us. She flew the jump, landed on the verge, got her hocks under her and slid down the bank rather like a small boy on a toboggan.

I rode her for two more days, catching her a couple of hard ones each time she went to sleep. By this time, her nature had changed so completely that I put a very good fourteen-year-old girl who was staying with us on her, gave them a week's schooling, then entered them for a novice hunter trial, which they won with ease.

The interesting thing about this was that in spite of this experience Donna retained her previous learning, and whenever she was given a weak rider she went back to her sleepy normal self.

So far we have been dealing with responses learned by reward or punishment. Now let us look at the effect of punishment on unlearned consummatory responses. By consummatory responses we mean responses that satisfy a primary need – for food, water, or sexual activity, so the response itself is the reward. The question is, how does punishment affect this kind of response? You might think that a consummatory response, being innate, would be hard to change.

However, experiments have shown that a rat receiving a shock at its food tray quickly learns to avoid the tray, no matter how hungry it is. It seems that it will starve itself to death rather than venture back to the food to see if the shock is still there. The reason for this lies in the principle of a conditioning stimulus: if the reward or punishment and the response are close enough together in time, learning is very fast even if the response being changed is an innate one. Even more dramatic results are seen when punishment for one consummatory response is combined with rewarding another.

An experiment conducted by Solomen in 1964 illustrates this. He found that puppies who were punished when they ate meat, but allowed to eat pellets without punishment, learnt very quickly to eat only pellets and to shun meat. The learned aversion proved to be so strong that the puppies were prepared to starve to death rather than eat meat.

This principle is important in dealing with horses, because we know that certain breeds of horses do certain things naturally, but we do not necessarily appreciate how easily this ability can be affected by conditioning. A race horse for instance will race naturally; because he is bred to race. He is bred not only for his ability to race, but also for his desire to race. However, if he is punished for racing – for instance he has a very hard, tight finish and the jockey carves him up to win the race – that horse may well cease to race, at least for a while, until the desire to race has been reinforced again by another experience. This theory of reinforcement is in fact important for innate abilities and desires just as it is for learned behaviour.

Let us take a thoroughbred, as an example. He has been bred to race and to gallop, and above all he needs excitement and movement. Yet if you take even a very highly strung thoroughbred and put him into a riding string, doing dull repetitive work, he will go off his food, lose condition and pine. There are a large number of thoroughbreds like this to be seen in riding schools. But on the other hand if you deprive a thoroughbred of galloping but at the same time replace this with something else, such as show jumping or dressage work, his desire to tear away flat-out may be changed to a desire to excel in competition, channelling ability, his liveliness and excitability in a completely different direction.

The problem of changing innate behaviour in horses reminds me of an old friend, Bryn Strap, so called because his christian name was Bryn, and his constant truancy from school led him to be beaten by his headmaster with great regularity: whenever Bryn failed to give a correct answer the headmaster would simply say 'Bryn, strap!' and try to administer some knowledge through his posterior. When I met him he was a tall, slim man of forty, immaculately shaved and dressed – which was all the more remarkable because he lived in a converted cow stall with a fire in the corner, the smoke from which escaped through a hole in the roof. The furniture consisted of a rickety table, two armchairs on the verge of collapse, a pile of blankets in the corner and no washing facilities nearer than the stream which ran below the house.

His manners were impeccable, his conversation extensive and he was as straight as a very bent corkscrew. He became a very great friend of mine, mainly because he could not only talk about any subject under the sun, but he was also an orator to make Lloyd George and Aneurin Bevan seem tongue-tied. This talent, combined with an ability to lie so that not only his listener but also he himself believed implicitly in what he was saying, would no doubt have taken him far, had he had a mind for public life.

As it was, he managed to sell me an orphaned two-year-old who made a coathanger look obese. After having her home and feeding her for a year she became a very nice little mare, about fourteen-two. The only thing you could not get her to do was to eat out of a bucket. If you emptied the feed on the floor she would lick up every morsel with great gusto; but put the feed into a bucket and she would die before she ate it. Sometime later I discovered the reason for this. She had been scavenging the local rubbish dump, trying to survive the winter before we bought her, when she had got an old galvanized bucket jammed over her nose. This had remained jammed like this for a matter of four or five days, until someone had managed to catch her and take it off. This is a clear example of how consummatory behaviour can be extinguished.

Rostellan provides another illustration. Since he is a Welsh cob, his main gait is a very fast and rather high-actioned trot, but over the course of time we have changed his behaviour fundamentally. Not only does he enjoy both cantering and galloping, his motive to excel, the great pleasure he gets from competition, and his natural asset of mighty hindquarters have actually combined to make him into an extremely good jumper.

An extremely important point to be remembered with horses is that if a horse has a genetic ability to do a certain thing, it is essential that such ability can be harnessed to overcome physical deficiencies. And some abilities, if they do not suit your purpose, can also be channelled in a new direction: a horse's desire to buck can be channelled into a desire to jump, for instance. In fact it is often said that if a horse bucks he will make a good jumper, because the muscles used for bucking and jumping are the same, and the mental approach which makes him wish to dominate his rider is similar to the will to win.

If you take a horse that naturally dominates the herd and try to make him go at the back of a bunch of horses he may give you a very uncomfortable ride. But on the other hand this dominance and competitive ability can be channelled to make him into a very good horse in competition. Or it can be converted into a sense of responsibility, if he is being ridden by the person who is in charge of a string of horses, because he will then enjoy the position of authority that his rider gives him.

Chance, for instance, when we used to have people on our farm on riding holidays, always had to be twenty or thirty yards in front of the other horses. But if I rode him when I was taking a party out he was quite happy to be at the back of the field supervising the weak riders and the horses that were misbehaving. His mental dominance was being made use of.

My experience with Dart shows how a horse's natural ability can be combined with his mental desires to overcome physical deficiencies. He again was a registered Welsh cob so not a natural jumper. But he had not only Rostellan's joy in competition and need to excel, he also had an incredible desire to win. I used him almost entirely for hunter trials and cross-country events. After his second or third competition, he was never beaten. His ability to beat thoroughbreds across country on time was almost unbelievable. First his desire to jump meant that there was never any risk of him running out or refusing; and second his great ability meant that he went into every fence flat out, and where it was necessary to take the next fence at an angle, he could twist himself in midair so that when he landed he was in the correct position to take the next fence. I had great empathy with him, so that most of the time I rode with a loose rein, steering him using telepathy combined with a slight change of balance. I only ever rode him once carrying a stick, and that was in a schooling session. He had developed the habit of napping slightly towards home, when doing a flying change to the off side. It was only a very slight hesitation, but wishing to cure him of the habit before it had developed, I decided to school him carrying a whip in my left hand and tapping his shoulder at the first sign of hesitation. This I did once successfully. On the second occasion I touched him once with the stick on his shoulder, and got a hair-raising response. Dart changed without warning to the off side suddenly putting us in front of a seven-foot fence consisting of a three-foot-six bank with three-foot-six of pig wire and barbed wire on top. Dart went at it like a torpedo and attempted to fly the whole lot, which he nearly managed, but he just caught his knees on the barbed wire, and turned a somersault throwing me fifty yards into the next field. Fortunately very little damage was done – Dart scratched his knees slightly and I excavated a large hole in the ground. I never carried a stick on him again. Nevertheless, at the end of each hunter trial or cross-country competition, Dart would end up in a state of complete exhaustion.

Another way of showing how a horse's natural ability can be channelled to a new purpose would be to examine a traditional method of teaching a young horse to work in harness. The colt would be placed between two old horses, into a set of drags. When the carter said 'hold fast', the two old horses went forward, pulling the colt with them. In a very short time the colt would be responding to the carter's command at the same time as the older horses did. Because he was between two old horses he was unable to go faster than they did, and soon he was imitating his seniors, responding to commands and pulling as well as he was able to. But imitation is not the whole story. The colt was having his desire to expend his youthful energy and power satisfied, and at the same time channelled into a rewarding task. His life was made more interesting by learning from his elders and the carter. The interest and fuss made of him by the carter also satisfied a desire, and in due course created a previously non-existent motive to please.

In nearly all these examples, we have been dealing in one way or another with the horse's tendency to associate experiences, or feelings, and to learn from the association. The horse will associate certain places, certain clothes you wear and certain stable preparations with appropriate forms of activity. If you plait your horse up the night before hunting, he will associate being plaited up with hunting; if you always jump in the same place, he will associate that place with jumping.

Weeping Roger associated racing so strongly with fatigue that he would slop around the paddock at a racecourse before a race, looking as if he were in the last stages of exhaustion. He would even wander down to the start in a tottering, lethargic way. But once he got warmed up racing, it was a different story. He always finished like a bullet, and when he won he would show his triumph by dancing and bucking back to the paddock.

We measure a horse's intelligence by his ability to learn. But at times one cannot always rely on the speed at which he learns as a direct measure of intelligence, because he may be using all his ingenuity to avoid learning something. So one has to set up a structured test. One such test is based on random feeding times. Instead of feeding the horse at regular times, say at eight, one, six and ten, you divide his feeds so that you are feeding at odd times and odd intervals. You do this so that he isn't associating a particular time of day with being fed. Then you position his feed so that you can get to it without him seeing you. Whilst you are out of sight you give a signal, such as a blast from a whistle, and repeat this each time you feed him. If, after the fourth or fifth blast of the whistle, the horse associates being fed with the whistle, and shows that he expects food, after a day or two you change the signal – instead of blowing a whistle you use a bell. This time, by the second or third signal a clever horse should associate the change to the ring of the bell with getting his food. Provided the horse is a normal feeder, you can assess its intelligence by this method quite quickly.

On the first test a response after four blasts shows intelligence, five is normal and six is slow. In the second test, if he is associating the bell with food at the second signal he's intelligent, at the third signal he is of average intelligence and at the fourth signal he is of low intelligence.

A number of simple tests like this can be evolved to test a horse's intelligence and quickness to learn. But be careful that you adjust your standards according to the test you use. In the feeding tests, a response to the fifth signal on the first test would be normal. But if instead the test were based on hunting with hunting clothes, and you only hunted the horse once a week, the association would take a bit longer – seven or eight times perhaps. It is very important to remember that the motivation, response and reward should be very close together. In the feeding experiment, the whistle should be followed by the response, which should be immediately followed by the food, in that order.

It is important to assess your horse's intelligence accurately, so that you can adjust your teaching to his learning speed. If you teach a horse too quickly, you will confuse him, and thus make him awkward and bloody-minded. If on the other hand your teaching falls behind his learning ability, he will put his intelligence to other uses, such as learning avoidance techniques and making his work more interesting by devising methods of annoying you. The art is in pairing his intelligence with your teaching, so that his interest is centred entirely upon his work.

This may be monitored by observing his enjoyment and desire to excel. It should be remembered of course that a horse is not born with instant intelligence. Intelligence is something that develops: the yearling will be more intelligent than the foal, the two-year-old more intelligent than the yearling, and so on. Maximum intelligence is not reached until the horse is seven, eight or nine years old. The drawback here, of course, is that by the time the horse is between seven and nine, any bad habits that he has learned will be very difficult to erase, (a) because the habit will be ingrained; (b) because the horse will be using its intelligence to avoid relearning; and (c) since he is at his maximum strength his fighting ability will be at its highest.

So an important factor to take into account when teaching a horse, is the previous learning that he brings with him to the lesson. A horse, once he has been through the initial training period, never learns anything from scratch. His previous training and learning and that which he has acquired from other horses is built on. This effect is called the transfer of learning, and can either be a help or a hindrance. Where learning is detrimental it is called negative transfer.

We have carried out a large number of trials and experiments to evolve a set of rules that govern positive and negative transfer, and found that these depend on the similarity between the stimuli and responses present in previous training, and those involved in the current learning situation. Where the stimuli are dissimilar but the responses are identical, the direction of the transfer will be positive, but where the stimuli are identical and the responses dissimilar, the direction of the transfer will be negative. Where the stimuli are similar and the responses identical, you will get a very strong direction of transfer, and when the stimuli are identical and the responses similar you will get a less strong direction of transfer.

Thus, positive transfer occurs when a horse learns to make a similar response to different stimuli. Suppose you taught your horse to jump a fence to gain a reward. Then suppose you used your heels and a stick to drive him over a similar fence, that is you made him jump a fence to avoid punishment. Here you have two different stimuli, one reward and the other punishment, but the response required is the same in each case: to jump a fence. The horse will make use of his previous learning – to jump a fence for reward – in order to jump a fence and avoid punishment.

Negative transfer takes place when the horse learns to make dissimilar or opposite responses to similar stimuli. Suppose you have trained your horse to respond to your legs, say pressing him with your right leg to make him turn to the left. At a later date, when you are teaching him to do a half or a full pass, you will press him with your right leg to bring his quarters in so that his head is pointing to the right. Here you have an example of exactly the same stimulus being used as part of the signal to get the horse to do a directly opposite movement. In the first of these you are using your right leg to bring his head and forehind to the left, which means his hindquarters, to maintain balance, swing slightly to the right. In the second of these, using your left knee to keep his forehind straight, you are using your right leg to make him turn his hindquarters slightly to the right, to move the horse sideways at right angles to the direction he was going. The success of this negative transfer will obviously depend on the horse's understanding of the whole context within which he is working.

Where the response similarity is identical, the nearness of the new stimulus to the old determines the amount of transfer between the two. Thus in the feeding test we did for intelligence, where the first stimulus was the whistle, when you change that stimulus to another signal, the nearer your second signal is to the first, the quicker will be the response to the stimulus and therefore the quicker the learning.

Finally, where a similar but not identical response is required, positive transfer, i.e. a helpful transfer, will depend on the stimulus being as nearly identical as possible. Thus if you have been jumping rustic timber fences it will be comparatively easy to teach him to jump colour jumps, if the same aids are used. On the other hand when you get to a more advanced stage of show jumping, and precise pacing is required – which means that you are checking him instead of letting him jump the fences at the speed and in the way he enjoys – your stimuli differ, therefore your learning is much slower.

For the same reason it is easy to teach a horse to go from a trot to a canter, if the same aids are used as you used to teach him to go from a walk to a trot. If on the other hand you use completely different aids, it will be more difficult. For example, when you are teaching a horse to go from a walk to a trot, you may simply squeeze him with your legs each time his school-master (that is, the horse you are using to teach him by example) goes from a walk to a trot. But if then it becomes necessary to teach him to go from a trot to a canter when the school-master is not there, each time you squeeze him hard he bucks, it will be necessary to teach him to respond to your voice instead. You try to get him to canter when you click your tongue, since you don't want a battle with him. By clicking your tongue each time he goes into a canter naturally, he will learn to canter when you click your tongue. It will take some time, however. This is an example of teaching a horse to do the same thing (change pace) using dissimilar stimuli. But it is always quicker and easier to teach something new if you can make use of something the horse has learned already.

11: *Developing Your Horse's Memory: Spaced and Massed Practice*

There are basically two different methods of training a horse, though most people combine the two, in one way or another. One is to do your training over a very long period, and this is called spaced practice. Spaced practice in its extreme form is illustrated by the person who will spend months teaching a horse to accept the bit before mounting it.

The other method is to cram a great deal of teaching into a very short period and this is called massed practice. In its extreme form, this might mean taking an unschooled horse out before he is going to a dressage test and cramming all the necessary training into a four- or five-hour period.

Both have their advantages and disadvantages. Mass practice capitalizes on motivating the trainer, who knows he has a one-day event the following day. And it also has the great advantage that the horse will remember its lessons fresh – the curve of forgetting will be slight, between training and event.

On the other hand, since the horse has learned an awful lot in a very short time, that knowledge is very thinly implanted and will be quickly forgotten in the long term. Spaced practice has the undoubted advantage that the learning will be retained better.

There is an exception to this rule. Over a period of time, massed practice will have an advantage over spaced practice only when the thing being taught is similar to something the horse has already learnt. So it is much better to teach a horse to jump a number of jumps, one after the other, than to have him jumping one jump over and over again, until he jumps that one perfectly, before he is taken on to another, which would be the method in spaced practice training. Massed practice in this case would have the advantage because the knowledge he had learned when jumping the first fence would help him jump the others.

For most forms of training, however, a steady and meaningful programme is far better than trying to teach a horse too much at one time.

Another question of strategy in training a horse is whether to use whole or part learning : whether to teach the whole of the outline of a piece of work in one go, or whether to split it up into smaller parts. With whole learning, the horse sees the complete picture and then you perfect the details. In part learning you perfect each detail, so reaching the same end product. Your decision here will depend on whether or not what you are trying to teach a horse follows on from previous training, and on how large and complicated the task is. You can then teach the whole of a relatively straightforward task, especially if parts of it are already familiar. But if the thing you are trying to teach the horse is complicated, long and unfamiliar, then it must be split up into smaller parts.

We did an experiment in whole learning on Chico, a three-year-old grey thoroughbred stallion, of 16 hands. He had been backed for five minutes as a demonstration of gentling in July, but handled no more until at the end of October when I got on his back and rode him for twenty yards, mainly to ensure that there were no great problems. The following day I saddled him and got on him, and rode him a mile to the meet, only to find that the hunt was taking place elsewhere, so though Chico had a brief sight of hounds, which proved of great interest to him, my wife and I decided to go home and Chico had no hunting that day. I got on him next a fortnight later when hounds were meeting nearby, and followed for an hour or so, steadily walking and trotting. A month later he was ridden again. We had a mile's hack to the meet and followed as close to hounds as we could get, again for an hour. There was a good meet three weeks after this, and we arrived just after hounds had found. Fortunately they swung our way and we could follow them closely for about ten minutes to a quarter of an hour. They then crossed a very deep gorge where we couldn't follow, but we caught up with them and the rest of the field a quarter of an hour later. We had to follow a line of heavy iron gates which, like the fences they were in, were unjumpable. Chico was going so well now that I was able to gallop him past the field, jump off, open gates and let the field through, remount him and gallop on to the next gate, repeating the process half a dozen times. I ended the day following Rostellan over two two-foot banks.

A month later when we took him out for the fifth and final time, we were in a piece of forestry land where we had twenty or thirty schooling fences. Fortunately, hounds found at the far end of the jumps and after about half an hour, swung back along the line of the fences. I have a habit of getting away by myself with hounds and Chico and I were by ourselves, so I cantered him up to the first fence. He tried to stop, but he was going too fast, so he did the only thing he could think of, which was to arch his back and buck. We landed safely on the other side, with Chico now reaching for his bit to catch up with the hounds.

The next fence had a hole knocked in it, so he took an extra large stride and flew it like a hurdle. He was beginning to find this new game fun, and jumped the next eight fences with improving style and ability, the last being a three-foot high tree trunk which we'd turned into a schooling steeplechase fence by packing brush against it. Twenty yards off he measured the fence, shortened his stride, accelerating in the last couple of strides and took it in a manner that would have done Red Rum credit. Since he was by now tiring, I called it a day and we went home.

From this it can be seen that, using the whole learning method, I had in five lessons taught Chico to stop, turn, gallop past other horses, allow me to jump off and on him at will and jump up to three foot.

As an interesting footnote to this experiment, I then started working him every day. The first day he went well; the second he went sluggishly, the third he went like a pig and the fourth day he spent in trying to buck me off. It took a fortnight's steady work to get him back to the excellence he had shown out hunting. This is a fine illustration of how, when a horse can see that what he is doing leads to his own enjoyment he learns quickly, but when the enjoyment ceases, and he is, as far as he is concerned, carrying out pointless work, his whole motivation is changed and his learning ability is severely retarded.

An example of part learning, on the other hand, would be a strategy used to get a horse to the stage where he is capable of jumping a complicated cross-country course. You start by teaching him to jump a series of small fences. Then you teach him to jump bigger and more complicated fences, and so on. In this case you split his learning into several parts, first teaching him to jump a series of obstacles, then teaching him to jump three-foot-nine with five-foot spreads, finally teaching him to jump a series of obstacles in awkward places *and* at awkward angles. (These obstacles are higher and wider than he would meet in competition and should be jumped faster than he would in competition, so that his competition work is infinitely easier than anything he has done at home. This is most important.)

Any piece of complicated learning should be split into small but related parts, each part following and building on the previous one. For example, when teaching a dressage test you first teach him to walk and stop, then you teach him to walk in a circle, then you teach him the transition from a walk to a trot, and so on. This will be done in two- or three-minute spells during his general exercise. Only when he is doing each of these things separately, completely correctly, should you take him into a dressage arena and link them together to make the whole. In my opinion it is only after the horse has reached a certain competence and is motivated to please you and to excel in whatever he is doing, that dressage should be attempted. In this I am against the general opinion, but I feel that a much higher standard can be reached with high motivation than may be reached from repetitive and habit-forming training.

In other words, the training programme, the method of work, must be right. But it should be used to force the trainer and the horse to respond positively.

If you can make use of the horse's motivation so that he is

actively and positively responding to your training, your results will improve startlingly. But still your method must be correct. The problem is like that of the small boy at school: if he is enjoying the subject he is learning and is taught so that he remains interested and actively working at the subject, he will do well in any examination later. But if on the other hand, he is only doing the minimum amount of work to avoid punishment, he will do badly in examination.

In the same way with horses, when you show a horse a jump, if he carries you over the jump with his own euthusiasm, then your training method is right. If he goes over the jump with the minimum of effort required to avoid being clobbered, or if he refuses or runs out, then your training programme is wrong.

Besides the characteristics of the trainer and the horse and the strategy used to train the horse, the remaining factor to take into account in teaching a horse is the nature of the task you want him to learn. Some things are very easy to teach and the horse will learn them easily; sometimes the thing you are trying to teach is very difficult and he will have great difficulty in executing a movement or jumping a particular type of fence. When your training is becoming advanced, in other words, learning becomes more difficult for the horse. Here you can make use of the horse's ability to perceive difference, you make a virtue of the specialness of the task.

A good example is the dressage test number five. Most of this test is a trotting one, but two parts are canter work, which will be completely different from the others. Now this distinctiveness can be made use of, to implant the learning in your horse's memory. When teaching show jumping in the course you will have a double and a spread fence. These will be distinctive in the horse's mind from the rest of the course. Therefore, if time is taken to split off the spread fences and the double, in teaching the whole you can give special emphasis to that part. So the most difficult and outstanding part of the training can be fixed in the horse's mind. But the fences must stand out because they are different, and not because they are difficult.

When teaching a horse to jump a double or spread fence, you start, of course, with a small double or small spread, and gradually increase the height. If you are teaching your horse with a group of other horses, you would have the main group of horses standing on the other side of the double or spread, so that he is jumping towards them. If you are teaching him by yourself, you should always have the double or spread pointing towards home, so that he has the extra motivation to jump. You will then make sure that you finish by jumping this fence and he is praised for doing so. And because it was the last fence he jumped before going home, he will associate it not only with being different (and so standing out in his mind) but also with praise and going home. His approach to it will be positive, and not negative as it would be if he were going away from home or away from the other horses.

Another technique for dealing with teaching advanced and difficult tasks is associating them where possible with training that he has gone through before. A coloured post and rails that he had been using in training for the show ring can be associated with any other post and rails that he has jumped, and you can always use his previous training in this way if you have been schooling over rustic fences, as we do. In fact, if you can split your difficult training programme into those parts that stand out, and those parts that you can associate with something the horse has done before, you should end up with very little that is new and that doesn't stand out. This part would be learned separately.

How wonderful it would be if a horse could remember everything we taught it! Every one would be such a wonderful performer. And yet it would spoil half the pleasure of riding, since there would be so little that you could teach a horse after twelve months of training! But the fact is that a horse forgets most of what you teach it, and forgets it much too rapidly for most of us.

So let us look now at the problem of forgetting. First we need to see how much he forgets or remembers; and then what changes take place in remembering and forgetting; and finally, the causes of his forgetting. When we have just finished teaching a horse something, we can be reasonably certain that at that moment he has 100 per cent retention of what we have taught him. But take him out some days later, and he may have retained only 75 per cent of what you have taught him – he has forgotten 25 per cent of what he has learned. Months later, he will probably have forgotten 75 per cent and retained only 25 per cent.

If you could look on the horse's memory as a curve and could chart what the horse remembers and forgets, you would see that he starts off at 100 per cent remembering and no forgetting, but then the amount that he remembers will drop very sharply for a time, then the curve would slowly straighten out at about 20 per cent remembering and 80 per cent forgetting. This figure is by no means fixed, but it is reasonable to assume for working purposes that a horse will retain about 20 per cent of everything you teach him. On the other hand it is a very important point to bear in mind that a horse's memory retention is at its highest when the horse's motive for remembering is greatest. This means that work he enjoys and the things that give him most pleasure will be retained in his memory bank for a very long time (four or five years), whereas dull and repetitive work will be forgotten as soon as the acquired behaviour has been replaced by something else. Our trekking ponies, for instance, will follow in line, day after day, never deviating from their position, for the period of the trekking season. But give them two or three days' hunting, and they will be prancing about like fools, their behaviour completely unrecognizable. On the other hand memory retention of something pleasurable is shown by the almost invariably long memory that a horse will have for where the feed room is! I have had a horse that I had sold, and bought back after as long as three years, walk straight to the feed room door. He had remembered that for three years, when turned loose in the yard, but had forgotten how to rein back five days after I had taught him to do so.

A horse will remember pain, as I have already stressed, far longer than he will remember pleasure. He will also remember excitement for a long time, and he will remember people for a long time as well. But the length of time a horse remembers anything depends in part on how well taught he was, the vividness and distinctiveness of his memory and on how much interference there has been since we have taught him.

From the very early stages I taught Biddy to change legs at the canter when she changed direction. This is important for any cross country work, but when we came to do a dressage test where the counter canter was needed, it was extremely difficult to teach her *not* to change legs when she changed direction. What is more, not only did she have to learn something which was completely contrary to her previous teaching, but she also had to learn that this method of cantering, i.e. the counter canter, was applicable only to the dressage arena, and that in all other situations she had to change legs, either automatically on change of direction or when I shifted my body weight.

The method of teaching her was in itself comparatively simple. When a counter canter was required in the dressage arena, if she was leading with her near fore I gave her the signals to lead on her near fore again; this reinforced that she was already leading on that leg. I combined this signal with the telepathic communication I was able to achieve with her, and her counter canter was extremely good.

I also made use of this teaching in cross country. When she was wrong at a fence, by signalling her to change legs twice I could get her to lose half a stride so that her take-off position was correct. Using the flying change in quick succession is extremely useful in slowing a horse who is going too fast – in fact on occasion I have put a bolting horse on the ground by using a rapid change of direction, so that he got his legs tied in a reef knot and went head over heels. But this is to be recommended only in dire emergencies or with horses who need a sharp and severe lesson.

Such a horse was Pippet, who came to us in the late 1940s as unridable. As soon as her feet touched grass she was gone like a shot from a gun. The first thing I tried was letting her gallop herself out round a twenty-acre field. After four attempts at this I tried facing her at a five-foot fence with a ditch on either side. In spite of my efforts to stop her she flew the lot. I didn't repeat this experiment because jumping five-foot fences with yawning ditches on horses out of control is not my idea of a way of prolonging active life!

So the next thing I tried was galloping her into a bramble bush. This stopped her, and she learned quickly to stop when faced with a sufficiently large bramble bush, but there were not enough about to make this a viable system of regular control. I finally taught her to change legs on command, and then took her into a freshly ploughed field. As soon as she started carting me I made her change legs quickly, which brought her down at once. I quickly sat on her head and made her lie still for ten minutes. After repeating the process four or five times I had only to change legs once or twice and she would come back on to her hocks completely under control. From then on she progressed to become an easily controlled and well mannered hunter.

Interference in learning is very important and is extremely difficult to cure in a horse. One of my first successes in eliminating interference involved teaching a horse to jump obstacles, many years ago.

Harry, a farmer acquaintance of mine, had bought a big slashing weight-carrying hunter for 700 gns. On his first day's hunting, being unused to our rhine country, the horse galloped into four rhines in succession, nearly drowning his owner in the process, so I got the job of getting him jumping water. Teaching him to do this was quite easy. I decided to do it on Harry's farm, for two reasons. First, it gave me considerable opportunity to spend time in the company of Harry's daughter Mary, especially as I was careful to pick market days, farm sales and any other time Harry was unlikely to be at home. Second, it was potato planting time at home and this was an excellent opportunity to avoid the backbreaking work of planting potatoes – such considerations as these are vital when you are considering the strategy of gentling horses. The actual relearning technique was simple. Since the horse naturally was a bold and brave jumper and simply had to learn to jump water, it was simply a question of preparing the ground.

The first thing I did was to get two ten-foot sheets of galvanized iron and tap them on to a wooden frame. This I placed in a gateway, and put a low eighteen-inch rail in front of it. Then I took Harry's big bay three fields away from home, where he had two big five-foot fences, and he flew them both with a leap of pleasure. Then we came to the gateway with the galvanized on the other side of the rail. He popped over the rail, landing with a crash as all four feet hit the galvanized. This had brought us back to the yard. I took his bridle off, loosened his girth and left him with a small feed of oats while Mary and I went into the house for half an hour's conversation and refreshment. Here you can see the motivation of both horse and rider.

After half an hour we went back out and took him over the same three fences, except I removed the rail in front of the galvanized. Again he jumped the first two fences perfectly. As we came towards the third fence and home, I drove him on. This time he took off five feet back from the galvanized and he flew it like a bird. Again he was put back into his stable with a small feed whilst we shifted the galvanized into a hollow in the ground in a different field. Again he had two fly fences before the hollow, thus keeping his motive of enjoyment alive. He galloped straight onto the galvanized, the clatter again alarmed him; I swung him round and put him back at the same place. This time he flew it without a mistake. From then on we proceeded to school him over any hollow or dip in the ground, one of which always contained the galvanized, until he was jumping any dip in the ground he came across. At the end of three days' very pleasant holiday he was jumping any rhine or stretch of water I put him at, so unfortunately it was back to planting potatoes.

This is why we insist that it is so important for a horse to enjoy his work, and that if he can see a reason for doing something and can understand it, he will retain the learning much longer than learning he has acquired but can see no reason for. Even interference in learning – probably the thing that has the greatest bearing on forgetting in horses (in this case, the interference of the horse's reaction to water) – can be overcome in this way.

We have seen the example of changing legs at the canter interfering with counter canter work. This is the kind of thing that happens all the time in training horses. You may teach a horse to jump properly, and then you make a mistake. Instead of giving him his head as he jumps you jag his mouth when he lands. Do this three times, and he will stop jumping because of the interference of his recent learning – that is when he jumps his mouth is hurt. Relearning then becomes a long and painful business.

Negative motivation, as we have already seen, makes a far deeper impression and is remembered far longer than positive motivation, because fear is a far stronger motivation than greed or love. And boredom and lethargy tend to reinforce forgetfulness, while alertness and activity tend to promote retention of lessons.

Alertness, then, is essential. It is stimulated by variation, boredom by repetition, and while everyone will pay lipservice to the idea that learning is enhanced by variation and alertness, it is a fact most people expect to teach by repeating the same lesson over and over again. Yet a little thought devoted to varying the work even a little will pay enormous dividends.

The last and final reason for forgetting I will deal with is outside interference. Say you are teaching a horse one thing, and his attention is diverted towards something else, he will tend to forget very quickly. This would seem to be obvious to anybody, but riders tend when training a horse to forget that a horse has a very short period of concentration – probably only four or five minutes – so that no lesson should last for more than this period. If your lesson exceeds his attention span, the horse is bound to be subject to outside interference. If, on the other hand, the horse concentrates for four or five minutes and is then allowed to relax and do something else, and come back to the lesson for another four or five minutes, all is well. If you try to increase this level of concentration, his level of memory retention will inevitably diminish.

External sensory information, such as may intrude in your lesson, can also be made to work for you. It is of course used by a horse to learn and function in everyday life. Sensory information comes to him as sensation, perceived through eyes, ears or skin. It is the brightness of a light, the pitch of a tone or the pain of a pinprick. Sensations are all stuff from which he moulds his experience. And yet his experience is much more than a series of sensations. In everyday life he is always interpreting the sensory information he receives. He interprets a sequence of sounds as a motor car, or a large green object as a tree, a square object as his stable or a cold wet sensation as rain. This process of interpreting sensations is called perception.

The fact about all perception is that it is always converting sensory information into meaning. A large red mass of certain shape is a car, a series of pressure sensations on his neck is a fly walking across him, a noise in the distance is another horse whinnying. Perception of objects is partly learned, but the basic tendency to organize stimuli into meaning is an innate property of the horse's sensory organs and nervous system. His natural ability to perceive objects, to create meaning, is called his organizing tendency.

One main organizing tendency which horses share with human beings is the ability to separate figures from their background. For instance we see objects, such as trees, standing out from the background of a field. Or one particular sound will stand out from other sounds, a gunshot from the sound of birdsong. The horse can select something, and separate it from its surroundings, with his eyes, ears, sense of smell or taste.

Another organizing tendency in the perception of objects is the ability to group stimuli into a pattern. This pattern means a person, that pattern a bramble bush. And he will group similar things together, and identify one man and another.

This ability to identify remains constant even though he may see one man standing on the horizon looking as if he is two inches high, and the other standing next to him who is six foot. Or the voice of one horse may come from half a mile away, very faintly, while the sound of another horse is bellowing in his ear : he will recognize them both as horses' voices and group them together. Similarly, he will perceive as constant qualities such as brightness, darkness, heat and cold.

There is another level of perception which horses appear to use, which human beings perhaps only occasionally share. This is extra-sensory perception, which is the ability to sense things they can neither see, touch, smell or hear. In one experience, this means the moods and emotions of his near companions. He will perceive your own moods in this way, and if you are nervous going into a particular fence, he may well refuse; or if he senses any weakness in your riding, use this knowledge to play you up.

On the other hand I find that I can use this faculty in a horse to get him to perform above his normal ability in com-petition or in special situations. In my demeanour and emotions I convey to the horse that the exceptional is required. I did this with Arctic Watch, the horse that ran away with me with great regularity and when racing would wrench the reins out of my hands and lead the field for as long as he was able. Once, the horse I had intended riding in the Llandeilo hunt team cross-country event went lame, and as a result Watch had to take part. My empathy with him was such that he perceived the difference in the situation and rose to the occasion. I dropped him in at the back of the team, from where I was able to observe how the rest of them were going. I found I could stop him when one of the other horses refused, to give it a lead over one of the fences, and turn him sharply to go back and catch one of the other horses that had deposited its rider. On the second half of the course I let him go on fast, to make up distance we had lost on the two leaders, by helping other people out of their difficulties. He allowed me to check him, turn and place him at each fence, catching up with the leaders long before they were finished.

This behaviour contrasted dramatically with his conduct the following Saturday when we were racing again. He pulled my arms out of their sockets to get into the lead, finishing third on this occasion. From this it can be seen that even though in both cases he was galloping with other horses, he was able to understand the difference in what was required of him, and this understanding depended entirely on the subtlety of his communication with me. Your understanding of a horse's perception, including his capacity for extra-sensory perception, can be made use of in any piece of learning, especially when teaching something new. In particular we use it in curing vices, in trying to make the horse change a bad habit for something we want him to do. And in curing each vice he has above all to perceive, after the initial treatment, that what we are trying to teach him is enjoyable.

12: *Eliminating Bad Habits: Stable Vices*

Some bad habits in horses are natural, but most of them are man-made. And unlearning a bad habit is one of the hardest tasks you can set your horse. It is a challenge to your teaching methods.

In this chapter, I propose to deal with the elimination of stable vices. These generally fall into two groups; those caused by boredom, and those caused partly by irritation and partly by the failure of the trainer to react to them before they develop.

Into the first group would fall crib biting. This clearly indicates a bored horse, hanging around a stable with nothing to see. On the other hand a horse that is difficult to groom, halter or saddle is frequently a horse with a thin skin that is easily irritated when he's being groomed, which in turn makes him kick or bite. If this is allowed to develop you end up with a horse that's almost impossible to touch. Both sets of vices are difficult to get rid of, and they require completely different methods.

Let us deal first with the group of vices caused by boredom. First it is important to eliminate the root cause, i.e. to see that the horse is no longer bored. This needs a two-pronged strategy. First, you increase his work and vary it, so making the horse more alert and interested when he is out. Then, when he is in the stable, leave the top half of the stable door open as much as possible. For most of the year he should be able to look out twenty-four hours a day, and even in bad weather he can wear extra rugs, even have a waterproof hood to keep his head and neck dry. He should also be put into a stable where he has maximum view of what is going on in the world outside – the rest of the farmyard, down at the house or, if you live in a town, the roads and houses nearby.

The main boredom stable vices are weaving, windsucking,

crib-biting and kicking the stable wall. In fact you can almost call these infectious diseases, because once one horse in the stable starts weaving or crib-biting you'll soon get others doing the same thing.

Weaving is most easily cured by turning the horse out for twelve months, but this isn't very often practicable since if you get a horse to ride or hunt you want to ride or hunt him, so you'll need to have him in. So the method I use to cure weaving is to make the horse's life as interesting as possible. Apart from leaving the stable door open, as I've already described, I give him toys to play with : I collect half a dozen Guinness cans, since these are usually plentiful about the place, and put some small stones in the bottom and then hang them on a series of strings across the door, so that when he looks out through the doorway there is just enough room for him to put his head through the middle. Now, if when he looks out he starts weaving from side to side, he's going to bash himself with the cans; but if he's bored he has only to move his head and he's got the tinkle, tinkle, tinkle of the cans to amuse him.

Within twenty-four hours of buying Salline at Llanybyther market, it became obvious that she was a very bad weaver and within another two days one of our younger horses had also started weaving. Daffyd in fact was quite easy to cure, simply by swearing at him. He quickly got the idea that weaving incurred our displeasure and probably meant that he got a stone thrown at him, though the stone, of course, never hit him, but caught the surround of the stable door a good thump. But Salline was a different case altogether. So we hung our empty Guinness cans around her door. She inspected them thoroughly, and then put out her head through the hole in the middle to see what was going on in the yard. She stood for four or five minutes, and then from habit she started weaving, first one way and then another, back again, back and forward. Then, when she was just getting into the swing of the weave, she caught herself on the face with a Guinness can. She jumped back into the stable, wondering what had happened. Ten minutes later she looked out again and started weaving again; again she hit herself. Over the following week she gradually weaved less and less, until at the end of ten days, instead of weaving, she would shake her head every five minutes or so just to hear the cans rattle. When she reached this stage we started to take down the Guinness cans. First we took down one on either side, then two, then we took them all down, by which time she was so engrossed in what was going on down at the house and in the other stables, among the cars going up the road and the sheep in the opposite field that she'd ceased to weave altogether.

The next stables vice is particularly annoying because it tends to cause so much damage. This is kicking, which appeals to horses that are stabled in wooden stalls because they find that if they lash back at the wall they get a good resounding thump. It tends to give the horse capped hocks, and the stable gets demolished very quickly. And since this often goes on especially at night, the thump, thump, thump from the stable wakes you up as well.

One cure is to hang a series of bags, half full of hay or straw, round the edge of the stable, so that when the horse kicks he's got a pretty fair chance of hitting one of them, which in turn bumps off the stable wall, giving him a bump on the backside. This is usually enough to distract the horse from bringing the loose box clattering down round his ears; though he may for a time continue to kick the bags, just to make them sway about. The worst stable kicker we ever had was old Cork Beg, and it took us a very long time trying to cure him before we hit upon this dodge.

Crib-biting, wind sucking and eating the stable are all manifestation of the same progression: the horse is bored, so first he bites at the stable door or his manger, just chewing it; then he really starts crib-biting, getting hold of the manger and scraping his teeth on it. Because at the same time he is arching his neck and inhaling air through his gullet, wind sucking then becomes a habit. This needs a great deal of time and patience to cure. So it is best to deal with the biting and chewing before it ever gets to the wind sucking stage.

The first step is comparatively simple. Since all these biting habits arise from boredom, and often happen mainly at night, you can put a piece of gorse bush, or the branch of a tree in the stable so that he's got something else to chew at, instead of the stable door.

I prefer this approach to the general recommended remedies, such as putting metal tops to all the surfaces that the horse is likely to eat, or using one of the greases and creams that are advertised, since none of these eliminate the root cause of the vice, which is boredom, though they may be useful in lessening the vice, and may be used – in fact have to be used in certain cases.

A determined crib-biter, however, may need more complicated treatment than the mere provision of a gorse bush. First, the important thing to do is to remove his wooden manger, or put him into a stable without a feed manger so that when you have fed him, you can take the feeding receptacle out of the stable. When you have done this, he will turn his attention to the top of the door. Here, in a minor case, you can simply put a series of tin tacks in a piece of cloth and fix this to the door so that he pricks his nose when he tries to bite the top of it. But he'll then look for something else to bite, so you proceed round the stable until you've eliminated everything that he's able to bite. Only then will some horses turn their attention to the gorse bush or branch that you've provided them with.

In very bad cases, the horse may simply tear the tin tacks off the top of the stable door, so you have to adopt much sterner methods. Here you want inch nails and a wooden batten, so that the inch nails stick out half an inch from the top of the stable door. You then do the same thing with any other surface he may want to chew.

Once the horse turns his attention to the branch you can remove the unpleasant nails or tin tacks, putting them carefully to one side in a place that you can remember so that you can put them back if he starts crib-biting again.

The same treatment will be used for the wind-sucker, except that here you may also need a six-inch wide piece of leather to strap round his neck under his chin to eliminate this very damaging vice.

Both these last two vices, wind-sucking and crib-biting, are extremely damaging to the horse's health. So both must be eliminated, and in the long run this can only be done by increasing the horse's interest in his work, combined with the toy in the stable and the obstacle to hurt him if indulges in his habit.

Here I must admit that when I allow my wife to read this book I shall remove these vital pages, just in case she tries them on me to stop me smoking! I don't relish the idea of picking up a Senior Service, putting it comfortably between my lips to take a drag and finding that I've driven a tin tack into my upper teeth.

There is an alternative way of tackling these vices, which is very effective if you have a neighbour who has an electric fence. This is simply to run an electric wire along the top of all likely surfaces, so that when the horse starts crib-biting or wind-sucking he gets an electric shock to his nose and mouth. This has the advantage over using tin tacks and nails that you are less likely to catch yourself on protruding points, but it is not always easy to find someone who will lend you an electric fencer.

The other problems I have called stable vices are not really so much stable vices as vices that happen to show themselves in the stable : kicking, biting, attacking you with the forelegs and squeezing you up against the wall of the stable. The first are forms of retaliation in response to annoyance; and the others, the really unpleasant ones, straightforward defensive aggression.

For example, if when you are grooming your horse he is tickled about his elbows, tummy or head, he will retaliate by snapping at you or flicking out with a hind leg. This is a simple reflex action of annoyance. He would do exactly the same thing to another horse who was annoying him.

On the other hand if when you enter the stable the horse comes at you with his teeth, this too is an avoidance reaction, but in this case the horse is trying to dominate you in one way or another. He may kick, present his backside towards you threatening to kick, or, in the most violent case, attack you with his front legs and his teeth.

To deal first with the pure avoidance and retaliatory reactions, there are a large number of remedies, any one of which may need to be used. Your general stance here is 'I know it's unpleasant, but you've damn well got to put up with it'. For example, when you're grooming under his armpits and he attempts to snap at you, a good hard slap on the nose will soon put him in his place. If when you're grooming his belly he leans against you, crushing you against the wall, a well directed fist into the ribs will immediately enable you to breath again. If he kicks out at you when you are grooming between his hind legs, again you slap or bring your knee up sharply, catching him in a vital and painful spot. These are pure retaliatory actions on your part for misbehaviour and if you catch a vital area you will have a well-behaved horse. The other place where horses sometimes dislike being groomed is around and between the ears. This one can be cured by tying him up extremely short so that he can't dance around, and standing on a bucket to brush between his ears. But watch it, because an extremely bad case will throw himself over backwards or sideways in temper. The only thing to do then is to resort to hand grooming the neck, ears and face. You simply run your hand up his mane, between his ears and down his face. At first you will find as soon as you get near his ears he will throw his head away from you. But you continue the process without a break until he allows you to do it without question. Then you repeat the process for several days, towards the end pausing to rub between the ears or the ticklish spot. When you have reached this stage successfully, you use a very soft small brush, and in time you will be able to graduate to a Dandy brush.

In all these cases where a horse is ticklish in a particular place, the first place you groom is that place. Not only do you groom him there first, you groom him thoroughly there until he is standing still. But it is most important to remember that if you're grooming in a ticklish spot it is only fair for you to respect it and use a soft or gentle brush, or your hand. After all, if you're ticklish about your feet and someone's tickling your toes, you too are going to go to any lengths to avoid having your toes tickled.

So far we have dealt with vices using the three basics of all horsemanship – determination, patience and understanding. But, from time to time you will get a horse that has dominated its previous owner and is really difficult to groom, or even to saddle. And here you may need some tricks. If he tends to lash out at you when you are grooming his belly, a simple trick is to put a surcingle around his middle and tie up a foreleg so that he is standing on only three legs. If he tries to kick you he goes flat on the ground and he won't attempt to do it again. After a period of time, you lower the tied leg by degrees. Only when he's got over his habit of trying to remove the vital parts of your body with the toe of his shoe and he'll stand still when you're grooming him, do you abandon the tying altogether.

If he tends to snap at you while you're grooming him or girthing him up, the simplest thing is to tie him up so that he can't reach you. In this way you hope to extinguish previous learning, that is, the habit. But if he does not forget, and still contrives to have a piece out of you, you have to extinguish the memory using fear. Extinguishing the memory by tying him up short works where the memory is not very deeply embedded. But where the habit is deeply ingrained within the horse, the only way to extinguish it is by using fear techniques. You can either hit him across the nose, or in severe cases punch his teeth in and really catch him one every time he tries to bite. This works *provided* you pay more attention to stopping him biting than you do to grooming him.

I found an extremely useful tool for doing this. It was an electric cattle goad, a device which, when you press it against an animal, emits an electric shock from the battery. I strapped it on to my arm so that whenever the horse swung his head round he touched the goad and got an electric shock to his face. This very quickly cured him of any idea that I was good eating. I tried it out on two horses and it was most effective. Unfortunately, the owner of the goad came and collected it, and since I didn't have one myself and couldn't afford to buy a new one, I went back to slapping my horses' faces.

As in all these methods of extinction where fear is used, it is absolutely essential that, having punched him, you should then make a fuss of him so that he understands that the ill treatment is for his actions and not because you don't like him. Whilst your immediate aim is to make him behave himself, your long-term aim is to make him like you and want to be with you. Handling a horse is rather like a love affair, it has its ups and downs. Things may be going right for you and the horse will be going superbly and you adore him, and then you have an argument and unless you're very careful, what starts off as a small disagreement can end as a severance of all affection.

Certainly before you can eradicate bad habits it is absolutely essential that there should be empathy and genuine affection between you and the horse. Also, biting and kicking when you are girthing or grooming or working in the stable is always caused by one of three things – boredom, irritation or fear. So the initial cause must be discovered before you can attempt to cure the vice. And once you have made the initial cure, you have still to proceed to the standard treatment of making him enjoy being groomed : that is, while you still tie him up to his manger, you put a small feed into him when you are grooming his ticklish places. So you are adding to the deterrent a pleasant experience, which he may come to associate with grooming itself; and if he is annoyed he can take out his temper by biting hard into his feed.

Another thing a horse may do in the stable is jerk his head up to stop you putting his bridle on, or grooming or clipping his head. This again is very annoying, and is one of the more difficult faults to overcome. You start off by rubbing your hand up his head and rubbing him in between the ears, holding on to the halter tightly. As he swings his head up out of your reach, you get a bucket or a box to stand on and do the same thing. You go on until you can get your hand between his ears and then you scratch the tickly place round the base of the ear, in the mane, between the ears and down under his chin. What you're doing is teaching him that what he is rebelling against can really be a pleasurable experience. Again a basin of nuts at the manger helps considerably in most cases. But the main thing is to keep at it and keep at it, maybe for an hour, maybe for two or three hours, until he allows you to do what you set out to do without objection.

Suppose a horse presents you with his backside, threatens to kick you or actually kicks at you as you go into the stable. This is caused by an ingrained fear, and it is the fear that must be eliminated. I do this quite simply to start off with, by putting him into his loose box without any straw. Then at lunchtime I walk in with a feed and a bucket of water. He turns his backside towards me and kicks out, so I take the bucket of water and the feed out again. I leave him there until the evening feed and walk in again; he will probably walk over to the corner of the stable, away from me, showing me a large amount of backside and the flash of an iron-shod hoof going past my ear. I take the water and feed out again.

By the morning, he will have been twenty-four hours without food and water and he'll be extremely hungry. One or two things then happen. Either, when I walk in with the food and water, he comes to the stable door and shoves his head into the bucket before I've got the door opened, and keeps it there while I open the door and walk in. Or he walks away. If he walks away on this, the third occasion, I stand in the open stable door. He can see the bucket of food and he's dying of hunger. He sees the bucket of water and feels as though he's spent six months in the Sahara desert. Slowly he'll turn right round – it might take half an hour or an hour – and come over to the feed. And in coming to the feed he comes to me. At no time in the case of a horse attempting to kick you do you approach the horse. He must come to you.

I have, on two occasions, had to leave the food inside and shut the stable door, standing outside until the horse comes over to the water and food. But as soon as he starts feeding or drinking I open the stable door. Within three or four days, when I go into the stable with his feed he comes to me, he has his feed and water and I make a fuss of him, and within a week or a fortnight, depending upon how deeply ingrained the vice is, when I walk into the stable the horse will remain with his head stuck over the stable door taking no notice of me whatsoever. Of course each time I go in I caress and pet him.

I have already, earlier in this book, dealt with a horse that cow-kicks at you with one leg. This is different from the horse that presents his backside or kicks you when you go into the loose box, since the cow-kick is an aggressive action as opposed to the defensive reaction of the hind-kick. To cure the cow-kicking horse you keep him tied, you will remember, so that he is unable to kick you when you enter the stable. Then you get in close to his body and make him kick, pushing his leg away from you so that he kicks at his other leg.

The final stable vice that has to be dealt with is one that is fortunately very uncommon, since most of its practitioners have already ended up in catmeat tins. This is attacking you with the teeth and front legs. The offenders are often very good stallions or mares who have not been handled a lot. If a horse attacks you with his front legs and teeth, what you do is respond in a similar manner – you roar at him at the top of your voice, lunge forward and punch him sharply on the end of the nose with a very quick uppercut. I've never yet found a horse that didn't retire immediately in surprise to the corner of his box at this, and a repetition of the treatment together with the strategy already described of going into his box only when you feed him, will in almost every case cure the horse completely. I say almost every case since I have had 100 per cent success, but I will admit that success does depend on the severity of your counter attack and on the person practising the cure. Having had a fight with the horse, you've got immediately to follow it by trying to re-establish an empathy with him. In the case of a horse that is kicking whenever you go near him, or biting or attacking you, you may do this by simply leaning over the stable door and talking to him in a sing-song voice.

One final cure that I used with great success on a horse we were sent to cure who would swing his head round and have a piece of you when you were out riding him, can also be used on horses who bite while being groomed. My technique was simple; when he bit me the second time I stopped him sharply, leant forward and fastened my teeth in his ear.

Charlie, when I was grooming him, used to take a large piece out of me. But after a course of this treatment, when I started grooming him he'd take the corner of my coat between his teeth and pull it back every time I tickled him. This to me was quite acceptable, because he himself had invented a method of telling me when I was getting too near the edge of his temper, and I would go off to groom another part and then come back to it. The horse was taking great care that he wouldn't hurt me, but he'd worked out a way to stop me tickling him unreasonably.

13: *Eliminating Bad Habits: Riding Vices*

Now that we have disposed in the previous chapter of stable vices, it is time to come on to the vices you encounter in riding your horse. All these vices are avoidance techniques of one kind or another. They stem from four causes: first pure bone idleness, second boredom, third malnutrition (which doesn't mean that the horse is starved, it just means that he is fed incorrectly), and fourth what you have taught him through your weakness and incompetence. The four causes may of course be interconnected.

Let us deal first with laziness. This can often be cured simply by changing the horse's feeding habits: what you're feeding him and the amount you're feeding him. He could be underfed. A weak, undernourished horse is clearly going to lack energy, so what you have to do is get him reasonably fit and well. Equally, he may be overfed. He will have very little desire to trot or canter if he is pig fat and off grass. And even a horse that looks in reasonable condition maybe being fed the wrong food.

I can illustrate what I mean by telling you about two horses we got last autumn. Swallow, my daughter's palamino, had been extremely ill with salmonella. He looked like a skeleton, and even when he had three parts recovered he was still extremely weak. When we backed him and rode him a little bit he was quiet and steady, but rather sluggish. Six months later, after having been fed in the meantime on horse and pony nuts and a limited amount of hay, he was jumping out of his skin and was an extremely good, if at times rather lively, ride.

Salline, on the other hand, was rather fat when we bought her. She was quite a good ride with a good walk, a good trot and easy canter, and was an easy and comfortable hunter. She stopped when we were cantering her with the merest touch of the rein, cantering easily to a fence and measuring each one precisely before jumping. But she was not very energetic. We put her on 14lb of horse and pony nuts a day up to Christmas, and as she got fitter she got a little livelier. She was still rather lazy, that is for a thoroughbred being got fit for racing, so at Christmas we followed what is our usual practice with point to point horses, and switched her from horse and pony to 18lb of racehorse nuts. Three months later she was doing an hour's exercise a day, two gallops a week and was racing fit. As soon as you started her at a canter she took hold and went straight into a good racing gallop.

These are only two examples of how altering a horse's feeding can change the whole nature and motivation of the horse. So with a sluggish or lazy horse, the first thing to do is take a very careful look at his feeding. Increase his hard protein-rich food, cut out soggy starchy food such as bran, flaked maize and sugar beet pulp, and go first on to straight horse and pony nuts, and, if that doesn't work, racehorse nuts. Restrict the amount of hay he gets. Then make sure he gets an hour's exercise a day, and you will very probably find that his whole nature will change. You will have of course to experiment to find exactly what food suits the individual horse. And your treatment will also vary according to your purpose. If you merely want to sharpen him up and make a good lively ride, you will not feed him in the same way as when you want him bouncing out of his skin and trying to jump every fence he meets. With the very large variety of foods on the market your choice is extremely wide.

If a horse tends to be lazy, fitness is the key word and that entails hard work on your part, because it involves not only proper feeding but regular and hard exercise. By that I mean an hour to an hour and a half exercise every day. And this work should be made as interesting and exciting as possible. Walking and galloping him with other horses will help. Jumping low fences will help, as will some variation in the place where you are working. You should not work the horse the same way every day: give him something different to look at every day, and when he is in his stable he should, as far as possible, always be able to look out and see what is going on.

One of the best ways to make a sluggish horse lively is to ride him in front of other horses, so that he get into the habit of being in front. And you will find very soon that when another horse comes up to pass him, he will quicken his stride to keep his pride of place. When you are cantering and galloping him, let him start racing the other horses.

Some twenty years ago a friend asked me to find something quiet, steady and grey for a friend of his, and so I went round all my contacts, and eventually found a 15.3 fat, lazy grey horse. He was quite good-looking except that he had a biggish head, so I bought him for a reasonable figure and the man who wanted him came and looked at him and rode him, hummed and hawed and hummed and hawed and said he'd let me know. About a week later he said he didn't want the horse because a friend was lending him one for the winter. I was landed with this grey fat thing that I'd christened the Zombie.

By this time I'd had him in for a fortnight and had been working him reasonably regularly. I'd groomed about a hundredweight of hair off his coat, mane and tail, and he was looking quite tidy. But I decided that the best way to sell him again was to hunt him for a bit and then sell him, so I put him on to 18lb of oats a day and cut his hay to about 6lb, which I gave him last thing at night with his last feed. By the time I got him out hunting for the first time he was beginning to lift his head a bit and was walking along quite well. It only took half a dozen hard swipes from my hunting crop to get him to trot, and if I really caught him a couple of hard ones he'd canter!

I spent most of the first day's hunting cursing and swearing as the rest of the field disappeared over the horizon, but towards the end of the day he started cantering when the other horses were cantering. I had him out the following Saturday and worked hell out of him in between. He had about three hours' exercise a day and all the oats he could eat: about 21 or 22lb a day. The following Saturday when we neared the meet, his head and tail came up and he started swaggering. When the rest of the field set sail he set sail after them, in fact he moved so quickly that I lost my reins and nearly disappeared over his tail! But we got together before the first fence, which he thundered into sounding like a herd of elephants. When we finally reached it he was going too fast to stop, so, seeing the other horses in front of him jump, he made a very creditable attempt, landed on the other side and pounded away across the field, every stride shaking my backbone and my teeth. But he was doing his best. Much to my surprise, at the end of half a mile we were still well in the middle of the field.

After a month's hunting he was eating 24lb of oats a day, and I stopped him there. I think he would have eaten more. He was really fit and strong and his jumping had improved to such an extent that he was always in the foremost group of horses all the time. From being a slow, lazy beginner's horse, he had become an enthusiast who was tearing for his bit as soon as he got on to grass, and he went flat out at every opportunity (admittedly his flat-out was slightly faster than the pace of a tortoise, but he was doing his best); and he turned into an extremely clever weight-carrying hunter. It didn't matter how awkward a fence was, he was changing legs, twisting and turning and jumping anything that came in front of him. The only thing he didn't like was a big spread, so I kept him out of water country, hunting him mainly with the Blackmore and Taunton Vale Foxhounds. At the end of the season he'd made such a name for himself that I sold him for more than ten times the amount I'd paid for him, and his new owner changed his name from Zombie to Springbok.

Just as correct feeding can change a lazy horse into a lively one, it can calm down a horse that is too lively and gassy. In this case, instead of increasing the protein and cutting out the carbohydrate, you put the horse on to a carbohydrate diet, giving him a minimum of oats but much greater quantities of bran, flaked maize and sugar beet pulp. And instead of making his exercise more exciting, you do the opposite. You still give him his half hour to an hour exercise every day, but it's all slow, steady walking, you cut out cantering and jumping completely until the horse has steadied down and has got to the stage where you want him. Once there, you regulate his food according to the work he's doing. If he's jumping, he'll need more oats: if he's jumping cross-country he'll need a great deal of oats, while if he's show jumping he'll need fewer. We once had a show jumper who had become almost impossible in the show jumping ring. As soon as he'd got into the ring he'd tear away at a rate of knots, getting over-excited and misjudging his fences. I only had him for a month, but as soon as he came I put him on a diet of flaked maize, bran, chopped mangel and chaff, filled his hay rack so that it was always full and gave him just half an hour or forty minutes' slow exercise each day. At the end of a fortnight he was going nice and steady at home, so I put him in the open class at one of the local gymkhanas and everyone sat back to enjoy the sight of me knocking fences in all directions. He flew into the first fence, very excited, but immediately afterwards he steadied down and did a clear round.

I took him home and kept him on slow steady work with no jumping whatsoever, plus his rather bulky diet. Then I took him out again on the Wednesday and finished second. He was out again several times in the next two weeks, performing very creditably, until his owner took him home.

The trouble had been simply that his owner had been schooling the horse, jumping him every day, feeding him like a king. But, this particular horse, to show jump successfully, needed slow steady work, a rather bulky diet and most of all, no jumping between shows.

There are some horses you can't steady down no matter what you do: they will always be too slow and steady, or too fast and excitable for you. If you find yourself with one of those you will have to change him for a horse that is better suited to your purposes. But before you do so, in fact in all cases where you are trying to change the motivation and nature of the horse, question whether you are giving the project enough patience or whether your riding ability is at fault. If you are not a strong enough horseman to sort him out yourself, it is infinitely better to get someone else to do so for you, or to change him for another horse.

Allied with excitability and laziness are two other vices, nappiness and rearing. Correcting nappiness is purely a question of patience and determination, possibly combined with punishment. Of course at the first sign of nappiness, you spank his bottom and tell him to behave himself. But if you have a horse who naps and won't go out of the gate, for example he half rears and refuses to budge at all, there are two things you can do. One is to sit on his back until he eventually goes forward the way you want: the other is to turn him round, back him four or five strides in the direction you want him to go, and then turn him again and see if he will go forward; if he doesn't you make him go back again. After a very few lessons he should get the idea that if he doesn't go forwards in the direction that you want him to go, he's got to go backwards in that direction.

In its more extreme form, nappiness becomes serious rearing. If a horse rears rather than go where you want him to, you keep his head pointing the way that you are going and keep niggling at him with your heels. Whatever you do you must not let him turn round and go home. Neither is it advisable to get someone else to come behind him with a hunting crop and catch him one on the backside, because you can't always have someone running about behind you with a hunting crop. When he goes up on his hind legs and tries to swing round, the answer is really a question of good steady patience. Again, backing the horse in the direction you want to go will very often help, but if his rearing gets worse you just turn him round, point him the way you want to go, and sit it out.

In both these cases it is absolutely essential that the horse goes where you want every time, every day, consistently for a week or two. And, as in all cases of curing vices, when he has done what he's been told to do, you must go out of your way to make the work that he's doing enjoyable.

In extremely bad cases of nappiness it is advisable to start by working him with another horse, because (a) if he's following another horse he's less likely to be nappy, (b) it is then quite simple to make the work enjoyable.

One of the most important things to remember in nappiness is that a horse never naps, or very rarely naps, when you are going towards home, so if you take him on a circular ride, so that you are always, even in an indirect and roundabout way, on your way home, you will find him easier than if you go out, turn round and then come back. As circumstances allow within your circular ride you can pick a place where he can do the work that he enjoys doing – maybe galloping up a string of low fences and maybe doing a little bit of dressage and schooling : I usually find that dressage tends to make the horse more nappy rather than cure the nappiness, but horses vary and you never know beforehand what he is going to enjoy.

One particularly nappy horse we had enjoyed splashing water in the stream. He'd stand in the middle of the stream with the water flowing round his knees and splash first with one foot and then with the other, bringing his foot as hard down on the water as he could. Then he would kick back his hind leg so that he was kicking water back behind him. He got so that even when it was time to go home to dinner, which was when I was beginning to get too wet, he was reluctant to go home. And within a very short time, as soon as you put the saddle on him, he'd bustle away down the road as fast as he could. When he was half way to the river he'd break into a trot, and then a canter and he'd take an almighty leap off the bank, landing in the water with a terrific splash. The nappiness had disappeared altogether. Then, once I'd got him away from the river – which I did with a couple of hefty belts in the belly with my heel – he'd walk home until he got half way, and again he'd break into a trot to get to his dinner. Such results can be expected in previously nappy horses once they get an enthusiasm for what they are doing. Nappiness, once cured, is one of the very few vices that is in very little danger of recurring.

A form of nappiness which isn't usually classed as such is refusing at a fence. This may be because the rider doesn't really want to jump anyway, and rides into the fence saying his or her prayers and wishing he had made sure that a doctor and two ambulances were waiting to take him to hospital afterwards. If you don't enjoy jumping, don't jump: it isn't necessary for every rider to do so, nor is it desirable, because not only will they be doing something that they don't want to do but they will also be ruining what could otherwise be a perfectly good horse.

Assuming however that the cause of the refusal is not in the mind and attitude of the person on the horse's back, it may be either the memory of an unpleasant experience in jumping, or pure damn laziness. He may have been jumped too much and too often, he may have attempted to jump a fence and hurt himself in doing so, or been hurt by his rider who has jagged his mouth or come down like a ton of bricks on the back of the saddle.

In any of these cases, the only way you can get him jumping freely and well again is by renewing his motivation and getting him to enjoy jumping. Here it is extremely important that the fences should not be high and that the horse should be allowed to jump them in the way that he enjoys jumping most. Some horses stop jumping because their rider has taken so much care to place them exactly at each fence that they've got sick and tired of being pulled together in a school, and so they say 'why the hell should I jump at all'. In this case you allow the horse to breeze into his fences at a good strong gallop, and fly them. Make sure of course that the fences are easily knocked down poles or brush fences so that if his pleasure in jumping is in demolishing the fences, or if he's wrong at one, he will not hurt himself. It is of course also important that he should jump with other horses, or if this isn't possible that all his jumping should be towards home.

Someone once brought a horse up to me because he wanted to school it over my fences, and he also wanted me to tell him what was going wrong to make the horse refuse. He cantered into the first fence very slowly, placed his horse beautifully and his horse popped over it. The next fence was a little higher, 3 foot, and again he cantered in slowly, placed it perfectly and jumped it. The third fence was three-foot-six high with a four-foot spread and again he placed his horse, cantering in superbly, but at the last second the horse stopped. He didn't, and made a beautiful jump, clearing the fence but leaving the horse on the other side.

He got back on, picked up his stick and caught the horse four or five clouts going away from the fence and four or five more going back into it. The horse climbed the fence somehow, leaving his hind legs on the second half of the double, and after that he refused to jump anything else. The rider asked my advice as to what he was doing wrong.

I simply got on the horse and took him at a fairish bat round the easiest part of the course, then came up at the sixth or seventh fence, still going at a good pace, into the fence he'd had trouble with before. Three strides out I drove my heels into the horse's ribs, one – two – three with each stride, increasing the pressure with my heels until he'd taken off. He rocketed over the parallel, heaved the reins out of my hand and away over the rest of the course, ending up with the four-foot gate followed by the four-foot shark's teeth fence.

In the course of the round he'd rattled four or five fences with his feet, but he pulled up blowing smoke like a dragon, with his head and tail stuck up in the air, and dancing around all over the place. So I got off and led him round for five minutes to cool down before I had a word with his owner, who couldn't believe the fences his horse had jumped. Neither could he understand how the horse had jumped all the fences at the pace we were going without being placed correctly. The answer to both questions was quite simple. The horse was enjoying himself, he was enjoying galloping and he was enjoying flying his fences; and whilst it appeared that I wasn't placing him, I was actually checking him three or four strides out so that he was going at an extended stride, or a slightly shortened stride, into each fence. He was a horse of scope and boldness who could jump the fixed timber fences with great ease, so it was merely a question of putting in a slightly bigger jump if he was wrong at his fence.

This method of jumping of course is not ideal for show jumping, but it is extremely effective going over cross country courses, hunting or racing. And even for show jumping, it is certainly more effective than being over careful in the placing of your horse so that the horse ceases to enjoy jumping at all. After all, since you're out of it if your horse starts refusing fences, it doesn't really matter if he knocks one or two down to begin with, and once you've got him jumping freely and easily and fast you can steady him, take your fences slower, and take more care over your placing before the jump.

One thing I don't like doing with a horse that is refusing his fences, is to catch him one with a stick, even though this may occasionally be necessary. With such a horse what I do is immediately to go back to a two-foot-six jump, then canter him into the fence, and if he refuses on no account do I allow him to turn his head away from the fence. I may back him three or four strides, but if I allow him to turn away from the fence without jumping it I am allowing him a victory, whereas if I keep him facing the fence, and make him jump it from three strides or a standstill (and any horse or pony can jump three-foot to three-foot-six with three strides) then the victory is mine. Sometimes, if you are by yourself, you have to tan him, but I much prefer to have someone standing ten or fifteen yards from the fence to pelt his bottom with small stones and curse him if he refuses, so that any pain and anger is coming not from the rider but from someone else. All the rider does is make him back three strides, urge him on vocally and with the heels, and praise and reward him afterwards.

I prefer to chuck a clod of earth or a small stone at the horse's backside, rather than hit him, for any form of nappiness, and this includes rearing.

There was a time when my wife's horse, Cuddles, took to refusing straw bale fences in competition. He would refuse them once, and then he would go back and jump them beautifully, which was a damn nuisance. We cured this very quickly. Whenever we cantered him into a straw bale fence, just as he got there, a stone would hit his bottom and I would shout 'Cuddles'.

This worked very well at home then at the first competition he again stopped at the bale fence. But there were two classes so when Leslie rode him in the second class I walked down to within twenty yards of the fence and, as he went past me, I said 'Cuddles' in a not very loud voice and he went into the fence and flew it. From this stage, when he responded to my cursing him he was able to transfer the response to my wife's simply scolding him as he approached any fence he was likely to refuse. You could see him tuck his tail in – which was usually stuck out at an angle – and prepare to fly over the fence, as if to avoid any stone that might be coming in his direction, even though we had long stopped throwing them. These stones were not very painful, but they were a great indignity.

The other common vice in jumping is running out. If a horse is likely to run out at a fence you need to collect him well beforehand to make him jump the fence well. If this doesn't work, you observe which side he runs – he will tend to run out consistently to one side or the other and, usually a horse will run out on his left if he is leading on his near fore and to the right if he's leading on his off fore. So you divide the fence mentally into three before you go. As you approach it, if the horse is leading on his near fore, you jump towards the right of the fence; if he's leading on his off fore you jump towards the left – in each case so that you have the wider part of the fence on the side that the horse is likely to run out.

In extremely bad cases you have to go back to jumping a foot or eighteen inches at a very slow pace, and keep him jumping slowly and collectedly except when he's in a lane where he can't run out, when you get him going at a proper pace. You progress from small fences to big fences, from trotting into them slowly to a fast trot, then to cantering in to them and eventually, after a lot of work and a lot of patience, you will get him jumping his fences fast and with efficiency.

It is a very common belief that refusing and running out are the same vice, simply because they are both used as avoidance techniques for jumping fences. And indeed some horses use both techniques to avoid jumping. Where this is the case, it is important that refusing be eliminated first, especially with horses who enjoy galloping fast and jumping. This is done by confining your jumping to a jumping lane or to fences so wide that a horse cannot run out. When the horse is enjoying his jumping so much that he is jumping everything at a slow, collected canter you can then concentrate on curing him of running out, which after your previous schooling, should be relatively easy.

Running out can take a very long time to eradicate, mainly because, for the horse, it is so well rewarded. As he goes to the left you go to the right and that's it, he's getting two rewards: a reward for avoiding the fence, and for getting you off. The reward for running out will then reinforce his motive to run out and so make it more difficult next time. His capacity to remember this particular vice is very high indeed. No matter how well you have cured him he'll try it on again every now and then, and you just have to go back and sort the problem out every time.

One more note on napping. An amazing number of nappy cases can be brought back to the ironmonger's: that is, they are caused by wrong bitting. In my belief there are basically only three bits to use – a plain snaffle, a vulcanite snaffle and a bar snaffle with egg butt ends. Anything else, if your horse is going well and correctly, is entirely unnecessary. Too much iron-mongery in your horse's mouth is merely an indication of your incompetence in riding him.

And remember that the pleasure that horses can get out of jumping is endless, and often quite spontaneous. I remember, out on a ride one day, passing a herd of four or five ponies who happened to be grazing a corner of the mountain near a piece of forest. As we passed the ponies just before we got to our jumping lane, three or four of the younger ones followed us and when our horses streamed up the jumping lane two of the ponies did the same. After that, whenever we were up that way those two ponies would follow us to the jumping lane and then gallop up the lane taking the fences as they came. These were wild and untouched ponies, and yet because they'd followed me that day having a gallop, and seen my horses jumping the fences, whenever they saw me going to the jumping lane afterwards they would follow me just because they enjoyed jumping. One form of nappiness I haven't yet dealt with is refusal by one horse to leave a group of other horses. This is usually not so much an evasive technique as an excess of herd instinct. The horse is a herd animal and breaking that herd instinct is extremely difficult. The only way to do this is with patience and perseverance, plus the occasional clout on the backside when he is being particularly bloody-minded. One of the things to remember here is that it is much easier to ride away from a group of horses at a right angle than at an acute angle. If you ride him away at an acute angle, he will tend to nap back towards the herd so that you are forced to pull his head round again and again. But if to begin with you teach him to ride directly away from the group, using your hand, heels and voice to direct him, his learning will be much more effective.

With a young horse we usually have him ride away from the group of horses led by an older and much more experienced horse, who already knows what to do. When he has learned quite easily to ride away from the main group with one horse leading him, we ride him away from the main group with the schoolmaster following him instead. At the third stage, the schoolmaster starts with him and goes a short way with him, then he stops. When the schoolmaster stops, we take great care to make sure that the pupil keeps going. The last and final step is to ride him away from the others directly.

If this is done there should be no difficulty in teaching a horse to leave the main group. This method is also useful when you are teaching jumping: you start off with the trainee following all the other horses, then you put him in the middle of the bunch, so that he goes third or fourth, then he goes second and then first, with the other horses jumping towards him. From there it is comparatively easy to get him to jump away from the main group, especially if you jump around a circle of jumps so that after he jumps away he finds himself swinging round back to the group.

The difficulty using other horses as an incentive in teaching is of course that it tends to multiply your difficulty when you're teaching him to go away from them. And in very bad cases of resistance, or with an old and bad mannered horse, it is often necessary to point your horse in the opposite direction and send the main group of horses away from you, before attempting to ride away from them. Over a period of time you diminish the distance between you and the other horses, until you can ride directly away from the main group.

14: *Rearing, Bolting and Bucking*

I have spent a considerable time on nappiness, since it is the most common and unnecessary vice among horses, and is tolerated by altogether too many riders through sheer weakness and incompetence. But we now come to more serious vices, which in their most severe forms can make a horse virtually unridable. The first point to remember, however, is that all these vices start off as very mild conditions. For example a nappy horse may go up a little on his hind legs, then more, and unless he is stopped at this stage you end up eventually with a horse who goes straight up and over backwards. To deal with all forms of rearing, from slight nappiness to the horse that throws himself straight over backwards, the same principle applies: that no horse will rear – or it is extremely unusual for a horse to rear – unless pressure is applied to the bit. So if you have your horse on a reasonably slack rein and his head is free, he is most unlikely to rear.

Rearing is one of the most dangerous vices because, when carried to its extreme not only do you hit the ground but you hit the ground with ten hundredweight of horse on top of you.

Years ago I had a horse who reared particularly badly, and I was consulting a friend, who knew a lot about horses, about it. And he came out with the opinion 'the trouble with that beggar is that you talk to too many women'. He wouldn't elaborate any further, so I went home and thought about this, and I realized that the horse only reared when I'd stopped to chat to one of my girlfriends. What was making him rear was the fact that I was making him stand still, and he was impatient to go on. Every time he tried to go on I checked him. At first he had reacted by throwing his head up. Then he started coming up on his hind legs a little bit, and by the time I talked to my friend about him he was really going up on his hind legs.

I cured him by never making him stand still. If we were waiting outside a cover while hounds drew, instead of standing still and chatting up a bird I would walk him round and round in a circle. This didn't improve my love life but it completely cured the horse of rearing.

My advice to anyone who has a really bad rearer is, first of all, flog all the ironmongery that you've got in his mouth to the local pub to decorate the bar with, and use the money to buy a plain rubber or vulcanite snaffle, which is the softest bit possible. Then never make him stand still. Keep him walking, walking, walking, walking. And never take him away from the other horses, because as soon as you do he'll start napping, which means that he'll go up on his hind legs again. After a month or so of this treatment, when he has stopped rearing completely and has no evident desire to rear, and his urge to walk round in circles has been considerably diminished because you've never allowed him to stand still, you can start letting him stand for a minute or two when he wants to. When he stands for a minute or two when *he* wants to, you can make him stand for a minute or two when *you* want to; and so you can extend the period until he will stand quite quietly.

Since the rearer is a horse of a particular psychological make up, with a very highly developed movement motivation, you will never be able to diminish his desire for movement for very long. You can however change his feeding to a high carbohydrate and a low protein diet.

If he rears when leaving other horses, your treatment is exactly the same as for nappiness: you first ride away with another horse, and then you go first, with your accompanying horse, and by degrees come to ride away from the field alone, without rearing. This again is a painstaking process because you are probably dealing with a deeply ingrained vice which has to be eliminated and forgotten over time. Long steady walking again helps in curing a horse who wants to rear, while faster work and jumping increase the vice. Only once you have cured him can you change his work so that he goes on to enjoy the things he enjoys doing.

When you get a horse that actually goes up on his hind legs and comes over backwards, there are two ways of curing him. In the first, as he gets to the point of overbalance, you slip off over his tail, let go of the reins and grab hold of the pommel of the saddle. Then as he goes down you come back up over his tail and into the saddle again. This is the one I use myself and I find it the most successful method of all. But split-second timing and athletic ability are necessary if you are to do it successfully.

There is an easier method. As the horse rears you slide off the saddle sideways, and as he comes down you vault back on the saddle. Again this takes athletic ability, but people who are curing horses who go up on their hind legs and throw themselves over backwards need a pretty high level of physical fitness anyway. Vaulting on and off a horse is a very minor accomplishment compared with some of the things you have to do when curing a bad habit in a horse.

With horses who go up on their hind legs when you try to mount them, you merely trot the horses forward, leading him, or with someone else leading him; then swing into the saddle at the run and keep him trotting. The next stage is to vault on him when he is walking, and then to make him stand still for a second while you vault on. Only after this do you slip your foot into the stirrup and get on while he is standing still. Slowly you extend the period for which you make him stand.

Rearing is usually a vice of thoroughbreds and Arabs and other horses of high excitability and great desire for movement.

Another avoidance technique, which is fortunately uncommon but extremely annoying when you come across it, is that of the horse that lies down and rolls over on top of you. This is an easy vice to cure provided you go about it the right way. It occurs in two forms. In the first the horse simply collapses on the ground and lies down. Here you simply sit on top of him, light a cigarette and admire the scenery until he gets up again. As with all these very bad vices, extreme coolness of head and control of your temper are necessary. It's no damn good hitting a horse with a very bad vice with a stick. You can be quite sure that by the time you've got him, if he's rearing up and going over backwards or lying down and rolling on top of you, he's been flayed alive by experts to try to cure him, so you may just as well leave the stick at home and rely on your patience. Eventually if the horse is simply lying down he'll find the grass rather damp and he'll get up again. You may have to sit on him for twenty minutes, half an hour, an hour or even two hours, but he'll get up sooner or later, especially if you stop him eating the grass near his head, because he will be faced with the problem that he either gets up with you on his back or else he dies of starvation and thirst. After he's gone through this treatment two or three times he'll stop the habit.

The second form, where as well as lying down the horse rolls on top of you, is also not difficult to cure. In this case you need a very old saddle that won't come to much harm, and as soon as he lies down you take your feet out of the stirrups. Then when he rolls over, you step over his tummy so that, as he comes up the other side, he gets up under you, and you're still in the saddle. I've never had a horse that tried this trick more than two or three times. When he finds that lying down and rolling over doesn't get you out of the saddle, he decides that this particular avoidance technique is no use to him, and he puts his mind to finding something else that will dislodge you.

Allied to rearing techniques are bolting techniques. These are all extremely unpleasant, and may need a certain amount of cunning to cure. One of the most unpleasant versions is where a horse carts you and heads for the nearest tree to try to brush you off against it. This may be cured in a single lesson, but does need cunning. As soon as the horse is out of control you try vigorously to swing him away from the tree then at the last moment you pull in the opposite direction so that he gallops straight into the tree and catches his head a terrible swipe. I have never had a horse try this a second time. But it is a risky technique: you must time it right at the last moment; and in any case if the horse is going fast enough and hits the tree head on, he may knock himself out.

With all these serious vices the remedies tend to be desperate ones, because either you cure them or the horse goes straight for cats' meat. But it is also true that these horses with extreme vices are very intelligent horses, who if it is possible – and it always is *possible* – to cure them become outstanding mounts to work on.

The mistake that most people make when they're on a horse that has bolted out of control, is to try to stop him by heaving on the reins. Now there are several ways of stopping a bolting horse, but one of them is not by a steady pull on the rein. All he does is shut his lower jaw, lean on the bit and use the pressure on the reins to balance himself. So, the first thing you try when your horse carts you is to give him his head, and then give a very sharp pull on the reins, lifting your hands at the same time so that you bring his head up and his weight back on his hocks.

Once you've found that your horse tends to bolt, the best thing you can do towards a permanent cure is to pick the largest field you can find and take him into it. As soon as he starts carting, let go of the reins and say 'right boy, get on with it'. After he has carried you two or three times round the field, you start to drive him on, saying 'right, now then, you go on', and push him for another half circuit or so. Then you pull the reins and say 'whoa' and he'll stop. Instead of fighting against you, you have allowed him to have his gallop and then made him stop when you told him to. Continue this treatment, and by degrees you will find that he'll stop carting you off in the same manner, and come back under control.

If your horse, when he carts you, goes off with his head in the air, one of the things I find most effective is a draw rein with a plain bar snaffle. Ignore your normal rein and let him go a hundred or two hundred yards, then heave on the draw rein and say 'whoa', pulling hard so that you have twice the strength to use on his mouth. This technique brings the horse's head down sharply on to his chest and brings him into a collected position.

The bolter must be allowed to have a good gallop every now

and then when he wants one, because the reason that he has taken to bolting is that you have suppressed his motive to gallop to such an extent that he takes matters into his own hands.

A third method of curing a horse that bolts is to keep changing direction. You allow him to gallop flat out, and then you turn him sharply to the left, sharply to the right, left, right, left, right. Each time he comes back to you, but again each time before you turn him you give him a slack rein so that he has to balance himself in his gallop without leaning on the bit. You go across or round the outside of a field on a slack rein. Then you lean back hard on the reins, at the same time leaning to the left, turning and pulling him to the left. More galloping on slack rein, then the same to the right.

The final method I use to stop a horse bolting is to put up a series of three-foot-six solid timber fences. After he has galloped into one of these, possibly putting himself and you head over heels, you will find that as he comes to the timber fence he will check and put himself right, or stop completely. At the same time as he checks, allowing him to go in on a slackish rein but making sure that he's headed straight at the fence so that he can't swing round and run out, you check him yourself, saying 'whoa'; and lean back to bring him on to his hocks so that he jumps correctly. This allows you a chance to pick him up and steady him, and then you let him go away to the next fence. As he approaches it you check him again, say 'whoa', and very quickly he will place himself and start responding to your commands to steady and stop.

With horses that bolt, you want as far as possible to make sure that they go out of a trot into a canter when they are pointing away from home and away from other horses, so decreasing the motivation to bolt.

I used two of the above methods with Watch, whom I thoroughly enjoyed racing for two seasons. But when he arrived he was extremely bad at bolting. So on the one hand I allowed him to gallop until he was tired. I did this by galloping him uphill in a straight line for about a mile and a half, then when he tired I pushed him on for two or three hundred yards and then said 'whoa' and he'd stop.

In addition, on the racecourse, I used a draw rein to steady him every time he looked like getting away from me. It took me about two months to get him so that I could ride him at home and check him and steady him, and it took me over a year to reach the stage at racing when I could drop him in at the back of the field and he'd settle down and gallop there.

Bolting is a vice that it is possible to cure, and since you know that you're almost certainly dealing with a horse that, in the long run, will probably be one of the best you've ever ridden, provided you're able to ride him, it is well worth taking the time and trouble to cure him.

The final serious vice, and a very common one in a horse, is bucking. Straightforward bucking is comparatively simple: I've already described how you find a steep hill, point the horse up it, drive your heels hard into the tenderest part you can find and say 'right, go on, buck you beggar', and each time he comes down you drive your heels in again. Half way up the hill he will have had enough and will stop bucking, and each time he takes to bucking again you repeat the cure. A horse bucking up a hill in a straight line is very easy to sit on, and you will be encouraged by the knowledge that a horse that bucks is also a horse that can jump, because the developments of the muscles – the powerful back muscles and quarters – which predispose a horse to buck are the very ones required for any form of jumping.

There are however a large number of ways of bucking, if the vice is well ingrained, and I will deal with those in detail. One variation is bucking again and again in the same place, while turning in a circle to the left or the right. If you've got a horse that is really bucking hard in a circle, he will throw you off sideways. Here you use the basic up-the-hill technique except that you carry a good long dressage whip, and when you get your horse to the bottom of the hill, as you drive your heels in to make him buck, you bring the dressage whip sharply across his neck and cheek on the side he turns towards (usually the left) when he's bucking. This immediately makes him straighten his head, and you go on using the whip to keep him going in a straight line.

This sounds brutal, and it is very hard on the horse, but please remember that all these extreme methods are used only in circumstances where the only alternative to curing the vice, is a bullet.

One possible method, that I don't use because I happen to enjoy sitting on a horse that bucks and curing him, is to put a saddle on the horse's back, tie up one leg to the girth, then put a tight strap round the back of the saddle, take a long rein and let the horse buck like hell round in a circle. By this means you do get the worst of the buck out of the horse before you get on his back. But since bucking is an avoidance technique and he's only bucking against the saddle, you're not eliminating the avoidance technique, which has to be done in a long term cure.

Some horses have a habit of bucking in a straight line four or five times, then stopping and puts in two or three big ones in the same place. Since you usually find that such a horse bucks forward with speed, the one he gets in when he's stopped is the one that gets you off. The only way to cure this is again to use your dressage whip. As you make him buck you bring the whip down across his backside, not on the side or the shoulder, so that you're driving him forward and driving him fast, and you will find that he will buck only three or four times before he settles down and gallops. Here you're changing the motive to buck to the motive to gallop, and as soon as he has learned that once he starts bucking he's got to go forwards and keep going forwards, the vice can be cured. Also, you should put him as quickly as possible into a jumping lane and get him jumping up the lane after other horses, so that his need to buck is diverted into jumping and enjoying himself.

The other two forms of bucking, both of them very rare, are switching ends and sun fishing. Switching ends means that the horse when he bucks will turn round in mid air so that he is facing more or less in the opposite direction; and sun fishing means that the horse twists his hind legs round as he bucks so that you are sitting at an angle of forty-five degrees. I've only had one of each of these in my lifetime, among maybe 150 horses I've cured of bucking, and the only way I could cure either was with a cradle. A cradle is very simple to make. You cut a number of sticks exactly the length of the horse's neck. Usually I use eight tied at equal distances so that they fit round the horse's neck exactly. This means that the horse has to keep his neck straight with his head pointed out. The bottom two sticks reach from the horse's chest to his chin, so that he has to stick his head out or he's driving the stick into his chest. This makes it impossible for the horse to buck at all. You work him and work him and work him, and you take a couple of inches off the sticks every week so that at the end of about three months he's really got a loop of sticks round his neck that aren't doing anything. When you've got to this stage and he's enjoying being ridden and worked, and you've got him jumping, you have made a complete cure.

You can use this method with any horse that bucks, but I've found that it doesn't make for a very long-term cure. In other words it can't be used as the sole remedy. But it does enable you to change the horse's method of bucking. From sun fishing, switching ends and turning in circles he begins to buck in a straight line, and it is then much simpler to cure him.

Finally, do not forget that the sooner you can get a bucking horse to enjoy jumping, the sooner he will forget his vice, and the motive to jump will replace the motive to buck.

Over the last three chapters we have dealt with a variety of vices, most of them in an extreme form. The point is that if you understand how to cure a vice at its very worst, you have only to scale the treatment down to cure it in its minor forms. All vices are basically avoidance techniques: a horse who doesn't want to do something finds a way to avoid it. And in this sense the way to cure a horse that doesn't want to be caught but gallops across to the other side of the field is in principle the same as the technique for curing bucking. You first make him allow himself to be caught and then, as a second stage, you make him want to be caught. You turn the horse out into a one- or two-acre field and then just walk round and round after him until he is so tired and bored with walking that he allows you to catch him. Then you make him do this four or five times until you can catch him whenever you want to. In the second stage, when you are teaching him to want to be caught, you also feed him in the field so that he comes over to you when he sees you with a bucket and sticks his head in it. So you put your arm over his neck and he'll walk in, eating his feed out of the bucket as he goes.

You teach a horse to lead in exactly the same way – you lead him with the bucket and you have your arm over his neck. As the next stage, as you lead him with the bucket from the field, you slip a piece of string round his neck; then the end of the halter. Then you put the nosepiece of the halter over his head before he can get at the oats. And all the time you're catching him and leading him with the bucket, the halter being the incidental piece of the movement.

Within a very short time a horse that was completely un-catchable becomes one who wants to be caught and wants to come in. At times in fact I'm beset with the opposite problem: that the horses are too keen to be caught. I go up to the field and give a shout and they all come over hell for leather to get at the bucket, and they're fighting and kicking at each other to get a mouthful, so that I am in imminent danger of being tramped into the ground. This is where order and discipline come in. Over the course of four or five days I teach them that they are caught in a set order. Jack, who is usually first at the bucket, has his mouthful and I halter him and give him to some-one else to hold. I go back, catch Madam, halter her and so on. In due course the horses will come in that order to be caught, each one having his two or three mouthfuls out of the bucket, each one being petted on being caught, and order and discipline triumphs.

Thus using each piece of learning which we have talked

about in this book in its correct place and in the correct way, you will make teaching your horse far easier for yourself. By increasing his learning ability the horse will learn to do each task with much greater ease. At the same time the end target of each piece of learning must be seen by the horse, so that when he has reached goal 'a' he must be able to see goal 'b', which leads on to goal 'c', and so on. How you mix which parts of his learning ability, which you use most and in what order depends mainly, on what you want to achieve in the end, but you also have to bear in mind your own abilities and the horse's characteristics.

But be careful what you teach the horse because it can have unforeseen results. You may remember how Cuddles learned to open his stable door and come down to our front door, and knock. We were rash enough then to teach him to press down the handle of the front door, and come in to get his feed just on the porch, and then actually in the sitting-room doorway. This was extremely amusing and of course a talking point among visitors. Until one day we came home and saw to our surprise that the front door had been left open. I picked up the shopping and went in, only to find that I couldn't open the door into the sitting-room. Something seemed to be wedged against it. So we went round the back door and in through the kitchen, to find Cuddles stuck in the sitting-room unable to go forwards or backwards or to turn round. Eventually, and with difficulty I lugged half the furniture out of the kitchen, got him into the kitchen and proceeded to turn him there. Unfortunately, as he swung his hindquarters round, he burned his bottom on the Raeburn, gave a bound forward, landed on my foot and shot through the door back into the sitting-room again. He refused absolutely to go back, and it took a good quarter of an hour of heaving and shoving and persuading before we could manoeuvre him out.

15: *Choosing Your Own Horse*

By now, if you don't own a horse already, you will no doubt be itching to have a horse on which to use and practice all these skills. It is impossible to study, understand, observe and work a horse consistently, unless you own a horse yourself, because understanding a horse properly is a full time job. So the only thing for it is to buy one yourself. This usually means buying the cheapest horse possible.

Cheap horses fall roughly into three categories. In the first come the old horses with so many bumps and lumps on them that they look like cheap coat racks. My advice is to avoid those. In the second category come the horses whose previous owners have been unable to handle them. These are the horses I like, and if you have the ability (this does not necessarily mean experience), the time and the patience to work on such a horse, you may end up with something outstanding. This is the way I get most of my best horses. And since you will be buying the horse at killing price, if you are successful you will be saving a very good horse from death. If on the other hand, you find that you are unable to cope with him, you will have to sell him before he kills you. And finally, in the third category of horses that you can buy cheaply you will find the plain, weak, unbroken three-year-olds. These almost certainly are the best group of horses for the beginner to purchase from. Since they are thin and weak, your preliminary work both in riding and working will be comparatively easy. Secondly the reward in taking a plain, thin, weak three-year-old and turning him into a well mannered, fit and well turned out five-year-old, ready to take part in competition, is one of the great rewards of horsemanship. You will also be turning a very cheap animal into an expensive one.

The first thing to be sure about is the height that you want to buy. And having decided, you must know that just because someone tells you that a horse is 14.1, the horse is not necessarily 14.1. He's probably about 12.2. So invest in a tape measure, or a measuring stick, and measure the damn thing. Then knock half an inch off for the shoes, and half an inch off for bad luck, and you're somewhere around the correct height of the horse. This process will eliminate fifty per cent of the horses you've been offered.

The next step is to decide how much you can afford to keep your horse. Moorland ponies for instance are comparatively cheap to keep – at the time I am writing, you can keep a moorland pony for £3 to £5 a week, depending upon the pony and the time of the year. But a thoroughbred is going to cost you £10 to £12 a week to keep. Next, remember that the pony has to be able to carry your weight, so if you can afford to buy and feed only a small pony, that is a 13.2, you want a 13.2 stocky pony not a 13.2 with legs like matchsticks. And then, decide what you want to do with your horse; if you want to do serious competition for instance, you need something big enough, and strong enough to compete on.

So, having decided the amount you can afford, both to buy and feed the animal, and approximately the size and strength you need, your next decision is what kind of temperament you are looking for. This will have a great deal to do with your own natural ability and confidence as a horseman or horsewoman. You may indeed be a genius on a horse. But geniuses on horses are very few and far between, so have a session of severe self-criticism, and decide what you want to do, and what you can cope with. Do you want to do just a little gentle hacking, a little quiet dressage work? Or do you want to compete seriously show jumping, or one-day eventing? Do you really want to hunt, or go in for team cross-country? Each of these activities needs a different type of horse.

Finally, when you have decided the type of horse you want, there is one more thing you must remember: that ability and good looks do not necessarily go together, in fact they are unlikely to go together, and pretty colours are not going to make a good horse. If you want a horse that is going to look pretty, and mince along like a pimp in Piccadilly, that's fine, but don't expect him to turn into a superb one-day eventer or a show jumper. You may be extremely lucky and find that he will, but the chances are against you. If you took the best show jumpers, the best one-day eventers and, the best steeplechasers in the country and put them in a show class, they would make a very long line at the bottom end of the class. The judge wouldn't look at them. Ability is something that is within the horse, and is not apparent on the surface.

Colours on the other hand are not irrelevant. They can be an indication – and I emphasize the word indication – of the horse's temperament. Strong colours tend towards strong character. Weak colours tend towards weak character. Now I know that there are a large number of exceptions to this rule, so don't put too much trust in it. But palaminos, washy chestnuts and greys do tend to be weaker characters than dark chestnuts, bright bays (who tend to be excitable), dark dirty browns (who tend to be steady and honest) and blacks, who tend to be strong characters and very good horses indeed. I myself have a definite bias to blacks; I have had a great number of them, and almost without exception they have been good horses: admittedly awkward and bloody-minded but they were each strong charactered with a real determination to win.

Children's Ponies

The breed of horses also makes a difference. I have very little experience of the northern breeds, so I am not going to pass any comment on them except for the Shetland pony, of which I have had a certain amount of experience, and which I abhor as a children's pony, mainly because it is stocky and broad, which makes for a difficult ride. Exmoors can be superb and very often are, but they tend to be one-man horses. Dartmoors, depending on which part of the moor they come from, again tend to be men's horses rather than children's ponies. With all mountain ponies you must remember their history. To survive they had to live on whatever they could pick up off the mountain; then, at three or four years old, they were herded in and roughly broken, and expected straight away to pull a heavy load, carry a man all day shepherding, or go down the pit and haul ten hundredweight in a tram. Only the very toughest survived: powerful, compact ponies of 11 or 12 hands whose very bloody-mindedness was their greatest asset. The one thing that they were not bred for was carrying children. The odd one who had the temperament and kindness to carry a child was the exception. The same applies to the Welsh mountain pony: only about 20 per cent of them are suitable for children to ride. The rest of them are either too excitable, or too bloody-minded. The 20 per cent are the ones that stop, and look after their riders.

The New Forest pony falls into a different category, because he tends to be a bit bigger. My personal preference for a child's pony, no matter how small the child, is something around the 13.2 mark – by that I mean from 13 to 14 hands. I know a small child on this is going to appear to be very much over horsed, but ponies this size tend to have a much steadier and easier movement than the smaller ones, and a much more reasonable temperament. No doubt a number of people will dispute this view, but as a general rule I find the 13 to 13.2 brown pony more likely to make a good, and safe ride for the child. After all, you are going to trust your offspring to the care of the horse, and the pony has to look after the child.

The most important thing in a child's first pony then, is the animal's temperament and reliability. It doesn't matter if it's knock-kneed, cow hocked, and has a head as big as an omnibus, provided it's sweet, kind and gentle so that if the child crawls between its legs and pulls itself on by pulling the pony's ears, the pony will put up with it. And if the child falls off, the pony will stop, nuzzle it and dry its tears. As the child improves, of course, you can progress to better things.

Also in buying a 13.2, of course, you have the added advantage that you also have something that you can get on yourself. If he's being a bit awkward you can put him through a term of re-schooling and discipline, until he is only too glad to get

back to his owner, and a quiet easy life again! It doesn't sound
very kind to the pony, but it is useful for him to learn that if he
deposits his rider and misbehaves too much someone is going
to get on his back, and make him work very hard, under very
strict control for the next two or three weeks!

An Adult's Horse

Here, the breed of horse you want depends on the work you
want to do. If you want quiet hacking, and a certain amount of
competition, my personal choice is the Welsh cob and the Welsh
cob cross. This is a really good horse, who enjoys competition
and has an awful lot of character and brains; and by and
large, Welsh cobs have a pleasant and good disposition. But if
you want something that looks pretty, and just want to hack
around, you could choose an Arab or Arab cross. These are
the most beautiful of horses, but as far as competition and jumping are concerned, they are not particularly good. I dislike
Arabs, but that is a personal prejudice, and plenty of other
people dislike Welsh cobs and thoroughbreds.

For genuine and serious competition – show jumping, eventing, or indeed anything that involves jumping and galloping –
you have of course to get as near to the thoroughbred as
possible. But remember here that – as with the Arab – you are
buying a horse that is going to be expensive to keep and feed.
This I know to my cost, because I usually have four or five
thoroughbreds knocking about the place, each of whom costs at
least twice as much to feed and keep as I do myself. They go
like the devil, in cross country competitions and hunter trials,
winning just often enough to keep me interested. But the trouble
is that each one I get convinces me that he's good enough to
make Red Rum look like a donkey from the sands. Then, when
I run him and he doesn't make it, he then convinces me with
his excuses: the going was wrong, he was not fit enough, he
got crossed at a vital fence, so I race them again, and again.
But I get a lot of pleasure wandering up to the stable, chatting
with them, schooling them and dreaming of one day actually
winning the big race.

By now you should have decided on the size of horse, and the sort of horse you want. Now comes the most important part of all, which is finding one that not only suits you, and your purse, but with whom you have an empathy. There are three ways of doing this. First of all, if you are going to a riding school regularly, you may find a horse there who goes better for you than anyone else, and who suits you down to the ground. Then, if your pocket is big enough, you may be able to buy him. Or, you may have a friend with a horse who suits you. Or you may go to a dealer (or someone who has advertised a horse, and anyone who is selling a horse is a horse dealer) and buy a horse from him. If you do this, you should insist on either a fortnight's trial, or an agreement that if the horse doesn't suit you, the dealer will have it back. In either case produce a pen and paper at once and say 'I want that in writing', because, while I'm not saying that all horse dealers (including myself) are dishonest, we tend, when selling a horse, to promise more than we intend to, and then to forget all about it afterwards.

The third way of buying a horse, which is the way I buy mine, is at public auction and, provided you don't pay a stupid price (and here you want the advice of someone who knows about horses), you can do whatever you want with it, and if, at the end of the month, you decide that it's not what you want, you can take it to another auction – not the same one, because there will be a lot of people who recognize it – and resell it.

Whether buying from a friend, a dealer, an acquaintance, or at a public auction, once you have ridden and tried your horse, if you don't feel an empathy and oneness with it, leave it alone. Whatever happens you must have a bond between you and the horse you are working with. If you don't feel this bond, don't put it off, get rid of him and get something else that does suit you. He may not suit anybody else, everyone else may say how ugly he is, or how awkward he is, or how useless he is, but if he clicks with you, and will do anything you ask him to within his ability, his appearance is of only secondary importance.

On the other hand it is of vital importance that you should have the horse vetted to make sure that he is organically sound, and that his legs are in a reasonable state, to allow him to do what you want him to do.

One of the ugliest horses I ever had – he was probably the ugliest horse I had ever seen in my life – was the most superb hunter and cross country horse I have ever ridden. Dart was 14.1 h.h., roman nosed, flat sided, long backed and bad tempered. After three months' handling, he became an absolute pet. But at the time I put him in his first hunter trial, he had jumped only half a dozen fences in his life. I had three or four other horses competing in the trial, which was about four miles away, and these had been led on with my wife's and my daughter's horses. But I had some work to finish and was going to follow later in the car. Then the car wouldn't start. So I slapped a saddle on Dart and we went hell for leather, because I was late. Having got him there, I decided to put him in the novice. He went like a bomb, being beaten in the jump off by my best horse, Fanny, and one other only because I was unable to turn him fast enough round the corner into the fifth fence. Except by Fanny, he was never beaten in a cross country again. He was an absolutely superb hunter, though as far as dressage was concerned, he thought it a waste of time, and the two or three times I rode him show jumping, he did very well but tended to sacrifice speed for precision. Nevertheless he was placed in two out of the three competitions that I rode him in.

The highlight of his career was giving a solo demonstration at the Royal Welsh Show, demonstrating the versatility of the Welsh cob. This included changing legs and doing a figure of eight, in twice his own length, jumping four foot, doing full passes, counter cantering and half passes at the canter.

Another example of looks being deceptive, is Mandryka. Seeing him in the paddock, he looks a small weedy 15 hand screw, but he is without doubt one of the best lady's point-to-pointers in the history of racing, having won over forty races. And a third example is my beloved Spitfire. No man weighing eleven stone seven, would dream of buying a pony of 12.2 as a hunter, yet she is almost certainly the best hunter I have ever ridden. Without any hesitation, I could put her at a five foot iron gate, and know that she would jump it. She would never turn her head at a fence. We could jump four foot of barbed wire and know that we'd clear it. In fact I would jump fences on her that any other horse in the British Isles would find impossible.

On one occasion when I was hunting hounds on her, hounds crossed a very deep dingle (a dingle is a Welsh term for a sunken river). It was about twenty feet below the field that I was in, and a yard or so back from the dingle was a barbed wire fence, three foot six high. I trotted Spitfire into the barbed wire fence, she jumped it, turned in mid air, landed at right angles to it, walked along the edge to a sheep path, and trotted down the sheep path into the dingle. There was a tree lying between where we came down and the sheep path on the other side. She jumped that and walked up the far side, where there was another sheep fence, three foot six high with barbed wire on top. There was about five foot in front of the fence, which was just enough for her to stand facing it, at an angle. She edged back slightly, until one foot went over the edge of the dingle, so that she knew exactly where she was, two feet from the fence. She jumped, arching over it, turning in mid air so that she landed in the field, and went straight away after hounds. The whole operation hadn't taken more than three minutes. Now there isn't another horse in the British Isles that could have negotiated that particular obstacle. Shortly after that, someone told me that it wasn't fair on the horse, and that I was going to come to serious injury myself, negotiating obstacles like that, but I said 'Don't be bloody silly, I can jump anything on this pony'. The old cow must have heard what I had said, because the next fence she came to she raced straight into it, and put me on the ground in a cow pat.

Getting to Know Your Horse

Suppose now you've chosen your horse, you've got him home and you still find he suits you. The first thing you have to do is watch and study him, not just when you're riding, but when he's in his stable and you are handling him. You have to discover whether he is a horse that stands still, in the corner of his stable or looking out of the door all day; or whether he spends his time walking round and shifting his feet. You need to know whether he drinks frequently, taking small sips, or whether he drinks half a bucket in one go; whether he eats all and anything put in front of him, or picks and chooses bits of his food. You must know whether he needs movement and interest outside, or whether he spends half his day asleep. You want to know the places he enjoys being scratched, and where he's ticklish when he's being groomed. You also need to know how he communicates with you, and other horses. All these things take time to learn. And all of them have implications for how you treat your horse. For example, if he's a horse that stands still, he's one that you will need to exercise a great deal. And if he is a horse that tends to sleep at the back of his stall, you will have to provide the stimulation to make him the alert, responsive animal he should be.

It is important to almost any horse to be able to see what is going on, so that he is mentally stimulated all the time. Ideally, he should be able to see the road, so that he can watch the cars going up and down, and to see the house, so that he can see what you are doing, and to see other animals moving around – anything that is interesting to watch and do. The more things he can see, the greater will be his enjoyment, and his readiness to respond.

Having provided him with a stimulating environment, you now need to give some thought to developing his brain, and his personality. This is really a question of a proper balance between insisting on your way and letting him have his own. Part of the time when you are working him, he must of course do what you tell him, and when he's done well and done what you've told him, he is praised for it and made a great fuss of. Then for ten minutes to a quarter of an hour, you let him do what he wants to do. If he wants to throw a couple of bucks and a fart, he's allowed to; if he wants to amble along, heaving a few mouthfuls of grass out of a hedge, he's allowed to do that. But when he's had his recreation, you say 'Right, back to work, off we go again' and then you do what you want. And so, depending on the length of the lesson, you carry on alternating work and pleasure. At certain times, of course, the things he wants to do, and the things you want to do are the same, for example when you are out hunting following hounds. But this coming together is really the end product of everything else you have done; his dressage work, his jumping, if he enjoys jumping, in his playtime, and your mutual pleasure in the excitement of galloping. It's an old, old saying, 'all work and no play makes Jack a dull boy,' but it's especially true with horses.

There is no doubt that training any horse *can* be done by repetition and habit, but full enjoyment of the horse (unless the enjoyment you get is from people saying 'Oh, isn't he well schooled' and 'doesn't she ride beautifully') is not going to come from repetition training, it is going to come from empathy and oneness between horse and rider.

The first stage in establishing this empathy with the horse, is in studying how he communicates. Most people understand some of what their horses are saying – when they say 'where's my bloody breakfast', demanding food, or when they show annoyance, anger or affection. But you can comprehend much more than this if you watch, observe and understand everything that your horse is saying, and at the same time you work mentally to get in tune with the horse. You will know when you have got in tune with the horse; when you are able, say sixty per cent of the time, to ride with a slack rein and still go where you want to go, at the pace you want to go. His movement will be reflecting the state of *your* mind, so that you'll get to the stage where you don't need to touch the reins, or make any movement at all, to turn him to the left or right, stop him, move him forward at a walk or a trot or a canter. In fact, I got to the stage with Arctic Watch, when I could take him into a field where there were three fences in a row, jump two fences and then stop, turn and go back over the two fences, without either touching the reins or moving my body. In fact in the last race I rode him I had no power in my left hand since I had injured it the week before, and since Watch's method of running a race was to go off hell for leather from the start, I was very much in his power. He was a very strong and powerful horse, and from start to finish of the race I only had minimum contact with his mouth, so I was really racing him with a completely slack rein. And yet old Watch, who was eleven years old, set off in his normal pace in the field, dropped back to the leading horse, and went stride for stride with him for a circuit and a half. This was an open race and Watch wasn't really an open race horse, and finally we dropped back to third place – apart from the fact that the old fool went straight through the last fence, we should have finished third. But the point is that there was complete mental empathy between us : he knew that I couldn't ride him – I could stay on top of him, but I couldn't steady him, turn him or place him at his jumps. Yet, except for the last, he jumped every fence perfectly. I was steadying, pacing him and jumping the old man purely by mental control.

Now this is what empathy with a horse is all about. You feel so much at one that he will make up for your deficiencies, just as when he makes a mistake, he will expect you to put him right.

If you can get a horse like Watch, Spits or any one of a dozen horses, with which I have got to that complete stage of empathy, you've got something that is of a price beyond rubies. Sometimes you gain it instantaneously with a horse but with others it takes a year, eighteen months, or even two years, but when you have achieved it, then, to use racing parlance, you can put your betting boots on, because then you've got a horse that is going to give every last ounce he has. You are going to have a horse that, when you've fought with your wife and kicked the cat, phoned the bank manager and told him exactly what you think of him and he has told you to take your overdraft elsewhere, you can get on, and he'll take you out, and he'll give a little jiggle and he'll give a little buck, and you'll suddenly find he makes you laugh and by the time you get back home, you're in a frame of mind to tell your wife you're sorry and kiss her, stroke the cat and phone the bank manager and apologize.

This is what he can do for you – he can lift you when you're depressed, he will mirror your enthusiasm in competition, he will extend himself when necessary, and above all when you take him riding, each time will be a new joy.

16: *Training Your Own Horse*

This is the point where the hard work starts. So we shall go through a guided review of schooling a horse from the beginning, from training a green three-year-old, curing vices, to developing him for competition.

The most important thing that anyone riding or working horses must remember, is that the horses are not people. They don't think like people, they don't behave like people, and their reactions are not those of human beings. A horse does not work from logic, he works and develops from previous memory. For example, if he has jumped a two foot plain rail, he will retain that in his memory, and when he sees a two-foot coloured rail, he will associate it with the rail that he has already jumped, and jump it successfully. When you are doing dressage, the horse will know from previous experience when he is to halt.

Timing and rhythm are both important in working with horses, and one of the facts that we know is that the horse has approximately the same ability as the human being to judge time. This skill as a rule develops later than it does in people but it means that when you are working a horse he will learn the rhythm and timing in the same way as you will. This is of particular importance in both dressage and show jumping, when you will often find that when you need to stand stationary for four seconds in dressage, for instance, the horse is more accurate than you are.

Now that you have bought your horse, you will want to start working him. For simplicity I am going to assume that the horse is completely unbroken when you get it.

With an unbroken horse, the first thing you want to do is to study it and get to know it. If he's never been touched he is going to be scared stiff of human beings, which means that the first thing you have to do is get his confidence. This is quite simple to do because you will be feeding him, watering him, cleaning him out, and generally being with him half a dozen times a day, and walking past his stable door twenty or thirty times a day.

Once he has accepted you as a friend and likes you, you can start putting your hand on him. You will know the time to do this, because when you feed him first thing in the morning, he comes bustling over to the bucket, for his breakfast. You can put a hand on him steadily, quietly and gently – just touch him with the tips of your fingers, behind his shoulder, rub him gently and working up his back and side until you get to his neck.

When you get to his neck you pinch his mane between your fingers, scratching at the same time, to try to imitate the feeling he would get from another horse nibbling at his mane, in the field. Don't forget, this is a gesture among horses and he will accept it and accept you as a friend.

When you have got to this stage, and he has accepted you as a friend, which may take three or four days, or may take two or three weeks, you can start teaching him to lead. This does not mean slapping a halter on his head, and heaving on the rope. Teaching him to lead is quite simple. When you take his feed into him, you keep hold of the bucket. As soon as he starts eating, you walk away with the bucket so that he follows you to eat his breakfast. Once he's doing this, walking round the stable following you and the bucket, you open the stable door and, following exactly the same procedure (making sure of course that the front gate is shut first), you walk round the yard with him eating out of the bucket, always making quite sure that you end up back in the stable before he has finished his breakfast.

The next stage is when, instead of him following you with the bucket, you hold the bucket in front of you, and rest your arm across his neck, so that he walks along with his head in the bucket, and your arm over his neck. It is vital at this stage of gentling and in all the early stages, at no time to have a battle with your horse. If he says no, you step back and come back to him slowly, and gently. You keep doing this patiently, until he does what you want him to. Anger breeds anger, so if you get him upset, frightened or angry, you are going to have a difficult horse to handle.

When he is walking around quite happily with his head under your neck, you take a bit of baler twine, and loop it round his neck, still using the bucket, to induce him to go where you want. A very short time after that, you wander into the stable with a bit of baler twine, put it round his neck, and you will be able to lead him simply with a bit of string round his neck, because he's learned that walking with you is a pleasurable experience.

While you are doing this, of course, you don't leave him in the stable all day. I like to have my horses out by day and in by night, making quite sure that when they come in their feed is waiting for them, and before very long they will be waiting at the gate of the field to come in for supper. At this stage you will be putting a piece of baler twine round his neck in the field, and leading him into the stable with it.

The next stage is haltering him. Haltering is very straight-forward too. When you put the bucket of feed into him in the morning, you simply put the nose piece of the halter round the top of the bucket, so that he has to put his head through the halter, to get his head into the bucket. The whole time you are using the motivation of appeasing his hunger, to get him to want to do what you want him to do. When he puts his nose through the halter, it is a simple matter to buckle the top piece of the halter over his ears, so that he has got his head collar on. And then when you catch him in the evening, you alter the practice slightly, you bring a small feed for him, so that he has to put his head down into the bucket, through the nose piece of the halter. Again you put the halter on and lead him in with the halter.

When you have got to this stage, and he is following you quite happily on the end of the halter, you take him for a walk, wandering down the road with him leading slowly and gently, letting him pick a piece of grass here and there, so that he is enjoying learning. In fact he is learning without realizing it.

One of the pieces of research that we did in teaching a horse to lead, was with a yearling colt. During his first winter, he was in by night and out by day. Each day when he came in, he was always the last one following the other three. Quite by chance to begin with, I used to walk holding his tail. Within a month I could stop him, turn him left or right, make him walk forward or trot, simply by turning his tail, pulling it slightly, or flicking it forward. Then, at the end of about three months I wanted to halter him. I had a half hour battle to get a halter on him, and then he refused to lead, and after that it took two people a fortnight to teach him to lead with a halter. One person had to hold him by the halter, while the other gave him the signals he knew and understood, using his tail.

It is a completely erroneous belief, that a horse has to be led with a halter, and certainly in the early stages it is a mistake to go straight to leading him in this way, since it leads to a battle with the horse and puts the whole relationship on the wrong basis in the beginning. Similarly we avoid a battle when we are teaching a horse to tie, by tying him fast in a stall with a bar behind him. If he hangs back, he is pushing against the bar with his backside, and not fighting against the halter. By degrees we tighten the halter rope, and have the bar further back, until he stands when tied without any trouble at all.

When you are in the stable with him, as well as making a fuss of him, you make a habit of moving your hands all over his body, as if you are grooming him with your hands. You increase the pressure here and there, you tease out a tangle in his mane, or tail. Now you are teaching him to groom. If while you are doing this quietly, he doesn't take any notice, you can take a stable rubber, and night and morning you can rub him all over, tidying up his coat smoothly. From the stable rubber, you graduate to the soft body brush. Again each stage, each step is a small step, and the feelings he gets are pleasurable, so he knows that each new thing you are doing with him is increasing his pleasure and increasing his comfort. At this stage also, when you are grooming his far side you lean over and push him over sideways with your body. When you want him to get over, don't try to train him to move over on command, push him over, so that he gets used to being pushed around physically by you. When he's used to you leaning on top of him, get someone to give you a hand over, and still with the stable rubber or body brush in your hand, groom the far side of him whilst you are lying on his back.

The next stage is to get a bridle on him. This can be done in two ways, depending on your horse. If he's going quietly, open his mouth gently and slip the bit into his mouth, and the bridle over his head. He should take very little notice of this, but if you have difficulty, put the halter on, tie the bit to one side of the halter, slip the bit through his mouth, and tie it on to the other side.

Once you start doing any action you go on doing it, and on doing it, until he accepts it quietly without any trouble. If for example, you have difficulty putting the bridle on, you stay there putting the bridle on and off, on and off, until you can do so without difficulty.

Then, having got the bridle on, you put the saddle on. Here you treat him as if he is an old and quiet horse. You go in with the stirrups and girth flapping, put the girth over his back and slide the saddle on his back. Don't go in with the attitude that you are doing a big thing – you're doing something normal, and quite straightforward. By this stage, he should be so used to you pushing him round and handling him, that a flapping saddle is something to be looked at, because it's interesting, rather than something to be frightened about. It is advisable to have a feed in his bucket when you saddle him, but if he is frightened in any way by the saddle, he is certainly not ready for you to get on his back.

Having got the saddle on, you catch hold of the girth, and

you put one buckle on the end hole, then you put the second buckle on the second hole, and slowly girth it up, until you feel his muscles tense. Then you talk to him for five minutes until he's relaxed again, then you go on tightening one hole at a time, until you've got the saddle on absolutely firmly. You then put the reins behind the stirrups, and go away and leave him for half an hour, so that he can wander round the stable with the saddle on and the stirrup leathers flapping.

When you've got the saddle on, or when you tighten the girth, perhaps two or three per cent of horses will proceed to buck. Just stay there looking at him and talking to him, telling him that he's a stupid sod, and if he wants to buck he can buck. When he stops bucking you go over to him and lead him round. If he bucks again you let go of the bridle, and let him go on bucking. Where you have a horse that bucks when you saddle him tightly – that is a cold-backed horse – before getting on to him you should always take him out with a rope and lead him round, so that if he's going to buck against the saddle he does so whilst you are on the ground.

When your horse has got tired of bucking against the saddle, and found he can't buck it off, you lead him round again and back into the stable, then get someone to give you a leg up, and lean across the saddle. Again, if he tightens his muscles, you slip from the saddle until he's relaxed, then go up again, and go on doing this until he's completely relaxed, with you lying there. Then you take your arm, over his back and down the far side. You do this a dozen or fifteen times, until he's gone fast asleep again, and then you ease your leg over the saddle, until you're sitting on it without him taking any notice. Rub your hand down his neck, scratch his mane and all his ticklish places, and slip off again. Then you get on again, on and off, on and off until he's got used to it.

Now you are ready to take him outside. If possible you should have someone on an old, quiet horse to accompany you; but if this is not possible, someone to lead him. First place him so that he's facing into a corner, so that he can't move forward. Then get someone to give you a leg up, ease your leg over the saddle, and once he is relaxed, turn him round. The person on the old horse now walks away, and you follow him. When your horse has settled down nicely and quietly, you say 'whoa', the person on the old horse stops, and your horse stops too, because if he doesn't he's going to run into the other horse's backside. You click your tongue, the horse in front will walk on, and you will follow. You do this for a quarter of an hour or twenty minutes, preferably in a circle. Exactly the same system is followed if someone else is leading the horse. When you say 'whoa', they stop and the horse will stop. When you click your tongue, the person leading walks forward and he will walk forward. By the time you get home, when you say 'whoa' and touch the reins, he should be just about stopping.

As soon as he's walking forward when you click your tongue and squeeze lightly with your legs, and stopping when you say 'whoa' and touch the bit, then you go on to the next step, which is getting him to neck rein to the left or the right. To turn left, you carry the rein right across his neck, so that you have one rein lying across his neck, and the other one is right out. At the same time, you turn your whole body to the left, so that your left knee is pressing into the saddle in front, and your right leg is pressing into his side on the right. When he's crossed the road to the left, you turn your body the other way, bringing the reins across the neck, so that he neck reins to the right. The next thing is to teach him to turn left or turn right at the end of the road.

When you've got him walking gently, slowly and nicely, turning to the left, turning to the right, and neck reining, which should take only three or four days, you get the person on the other horse to trot four or five strides when you click your tongue. Again your horse will imitate the leading horse, trotting to catch him up. Let your trotting period increase from four or five strides, to ten or twenty yards, then to fifty yards. To start off with, you will do a slow sitting trot, and when he has settled to that you do a rising trot. By now you will be doing half an hour to forty minutes' work a day on him.

Find a nice straight piece of grass, and go from a trot into a canter. Here, very occasionally, you may find that he will put in a little buck, though this applies only to one horse in twenty. When we first canter we canter uphill anyway, not because we expect the horse to buck, but because if he does buck he will find bucking hard work, so he'll stop. Canter steadily up the hill, and by the time he gets to the top of the hill, he will be a bit puffed and tend to drop back to a trot. If you feel that he's going to drop back to a trot, you say a 'steady boy', pull on the reins and pull him back, so that he has made the transition from the canter to the trot, at your command. You treat each transition in fact in the same way : from a walk to a trot, from the trot to the canter and back again. You give the command, just as you feel that he is going to do the movement anyway.

Now you have a programme of straightforward walking, trotting, cantering, turning each day. You are in effect working on two things at once : his motivation; and his fitness. He is enjoying working and riding with other horses, enjoying imitating his elders, going to different places every day, and seeing different things. His interest is aroused, his motivation is increased. At the same time he is getting fitter, and to help him do so you increase his feeds. One of the things that I particularly abhor, is the practice of underfeeding horses when they are being broken − feeding them hay instead of corn. I always believe that if a horse is doing more work, he will need increased feeding. This may lead to a little extra trouble for you, but the important thing is that each new piece of work that the horse does at this stage he should enjoy doing.

The empathy and communication that I talked about in the last chapter, will be growing as you work. You will be getting closer and closer to the horse. Now when you are out working him you make him do a collected walk. Pull him together so that he's collected, not just slopping along any old how. You will do this for two or three minutes, half a dozen times during the ride, then you increase the length of time. Particularly before you make a transition from a walk to a trot or from a trot to a canter, or back to the walk, you pull him together and get his head in the correct position, so that the transition becomes a collected transition.

You do each thing as you feel you *can* do it. The important thing is that what you are doing with the horse, should be geared to the pace that the horse wants. You must never move too slowly, so that he gets bored with the work he is doing – ideally he wants to be doing something new each day so that he is keeping his mind active – nor too fast, expecting the horse to do more than he is ready to do.

Soon now you will be ready to start teaching him to jump. By that I don't mean that you build a series of five-foot barriers, and drive him into them hoping for the best. This is in fact the way a number of people train their horses to jump for point to pointing; they jump them over a couple of gorse bushes at home, put him into a maiden point-to-point with some fool like myself on his back, and you gallop into the first fence, and he may or may not scramble over it. And this actually works after a fashion, because after he's run in three or four point-to-points, the pleasure of racing with other horses does usually get him jumping. He learns by his mistakes, and by the bruises you collect. But this is a crude method for teaching horses to jump, rather like throwing someone in the deep end of a swimming pool.

I remember on one occasion arriving at a point-to-point to find that the owner that I rode for had a horse in the lorry which I had not seen before, belonging to his neighbour. It turned out that it was a little grey mare called Passing Cloud III, and he wanted me to ride it in the maiden. I stood in the paddock looking at the ill-kempt little weed, comparing it unfavourably with the superb sleekness and fitness of another newcomer, Four Ten, who subsequently won the Gold Cup. But I couldn't offend the owner, so up I went. I got a good start on the little mare, leading into the first fence. Twenty yards from it she slowed, looking with horror at the obstacle in front of her, but the twenty other horses were bustling past her. So she stopped at the bottom of the fence, took off like a helicopter, and somehow landed on all four feet on the other side. She was a game, gutsy little mare, and dashed away after the others at a rate of knots, adopting approximately the same technique in negotiating the next two fences. The fourth fence was the open ditch. Certain that we would come to grief here, I decided to do so in style. Three paces out I caught her a couple, so that she accelerated, and from pure good fortune she took off at the right place, and jumped it superbly. She must have been an extremely quick learner, because from then on she jumped beautifully. Since the rest of the field had been going like hell, trying to keep up with Four Ten, whilst I had been doing my best to keep my neck attached to my body, we had lost an awful lot of ground but I kept the little mare going easily at a reasonable gallop, and to my surprise, after we had gone about two miles, we started passing horses that had blown up, and whose jockeys had given up the ghost. Another half mile and we were passing the horses labouring towards the finish.

There were only four horses struggling after Four Ten, two of whom fell at the last fence, and going up the last hill the little mare sprinted past the remaining two, to finish second – though I must admit that by this time Four Ten was back in the unsaddling enclosure and his jockey was weighing in. Nevertheless the little mare won her maiden flat race a couple of weeks later, and a hunt race at the end of the season, so all in all she was a lucky ride for me. By a strange coincidence I also won that season on a very good horse called Passing Cloud IV, who was also a novice.

But by far the best way to teach a horse to jump, is in the course of your ordinary riding. You simply look out for a series of small ditches, maybe six or eight inches deep and a foot across, so that he can almost step across them. But when he sees that the horse you are with breaks into a trot and flies over, your horse too will break into a trot, come to the ditch, and either stumble over it or, more likely, arch himself and give an almighty leap. When you have finally discovered where the hell your saddle has got to, retrieved your reins and stopped, you tell him how clever he is and make a fuss of him. You do this four or five times, if possible over a different ditch each time.

You then progress to larger and larger ditches, until by the end of the ride, whenever he sees his companion pricking his ears, and starting to canter, he should be tearing after him, to join in the new game you have devised for him.

When he is jumping ditches happily and successfully, you go to the next stage of your schooling, which is finding a convenient verge eighteen inches or two feet higher than the side of the road. Again he follows his companion, popping up from the side of the road onto the verge. Following this, you teach him to jump from a higher piece of ground to a lower piece of ground, jumping down over a drop again eighteen inches to two feet.

Teaching him to jump banks like this, first up, then down, has the advantage that he has time to collect himself after doing the first half, before he does the second half, which is the drop. After half a dozen or a dozen little jumps, he is enjoying the new experience. So you start by jumping two-foot banks, still preferably from a trot; but some horses won't give you much option, and when they see their companion cantering into the bank, they will tear the reins out of your hands, and go cantering into it too.

Each of these lessons, of course, is interspersed with normal schooling. By now you should be attempting flying changes and figures of eight. These are very important when you are schooling a horse to jump, since when jumping fixed fences, one of the essential methods of putting your horse right, is to do a single or double change of leg.

Teaching a horse to do a flying change is quite simple. If your horse naturally leads with his near fore at the canter, you get him going into the canter so that he leads with his off fore, that is the leg that he does not naturally lead with. You then canter down the field, check him and turn him sharply to the near side, and in most cases he will quite naturally change legs. You use your body, of course, leaning back when you check him, and then throwing your weight onto his near fore, so that he has to adjust his balance by changing legs. When he is changing on command, from his unnatural to his natural leading leg, you will reverse the process. Best results will be obtained by working him at the canter on his off fore for three or four days beforehand, then simply starting him on his near fore, checking and turning at right angles, using your body as a counter weight.

You will notice that most of the things that I am advocating go against the dressage methods, but for hunting, cross country, and any form of speed jumping competition, if you can convey your wishes to your horse by merely shifting your body weight, you will not only be speeding up communication, but you will also be assisting his balance of movement.

When your horse is jumping ditches and banks with enjoyment and fluency, you graduate to fly fences. These should preferably be jumped towards home, and I prefer loose poles of brush, eighteen inches to two feet high. If he makes a mess of one of them, he can gallop straight through it without hurting himself. When he's tearing over these, dancing and snorting as he comes to his jumping place, you can put a rail six inches above the brush. As he progresses the rail gets higher, and when he has got to about three foot, you start him jumping straight post and rails.

When this stage has been successfully completed, and he is looking forward to his jumping as a reward for his schooling work, you can start introducing spreads. To start off with, your spread will need to be two foot nine inches high, and two foot six inches wide. This of course gets bigger and wider, until his spreads are three foot six inches to four foot apart, and his normal post and rails is three foot six to three foot nine inches. After this you introduce combinations, again dropping your fences and spread widths to below the three foot mark at the beginning.

At the same time as you are schooling your horse over fixed timber fences, you can also be schooling him to jump water. This is done by first picking a nice sunny day, riding him down to the nearest river, and letting him play in it for ten minutes. Then you ride him down the river, until you come to a nice bank to jump out from, and you jump him back in the water, so that he enjoys the splash. When you have got him enjoying jumping in and out of the river, you place a low post and rails, so that he jumps the post and rails into the river. You can put various obstacles in the river for him to jump, each time making sure that he enjoys playing in the water.

You get him to jump fences at right angles to each other, by first jumping him over one fence, and then turning him to jump another at right angles twenty or thirty yards away. By degrees, you bring them closer and closer together, until he can jump a three foot six post and a rail, turn on his hocks and jump out over another.

By this time you will have taught him to jump almost any fence that you are likely to meet, either when you are out hunting or across country. The three things that you have to remember are, first, that you have to keep within the capabilities of the horse – if you have any trouble you go back to a lower and easier fence; second you have to see that he learns to place himself and jump naturally but, at the same time, responds to the movements of your body, so that you can put him right where necessary; and third, and most important, that he enjoys his jumping and looks upon it as a reward.

There is one final fence I would say that it is necessary to learn to jump, although this is not part of normal schooling. This is barbed wire. Barbed wire is much feared in this country, and yet in Australia and New Zealand it is considered to be a normal fence, and it is really quite easy to teach your horse to jump it. You stretch a strand of plain wire over a rail at three foot, or three foot three inches. Trot him into the fence (don't canter him) and he will probably clear the lot. He may hit the wire at first, which will knock his feet back, but since he's used to jumping a three foot rail, he will clear the rail anyway, and soon learn to clear the wire as well. Then, when he is jumping this quite happily and easily, you remove the rail and put a second strand of wire below the first.

Each time before jumping wire, especially on any young horse, you walk him up to it so that he can feel the height of the wire on his chest. And you click your tongue when you want him to take off, and you say 'hup', encouraging him to jump on command. This training to jump on command is a very important part of training over wire, and you can get some horses to the point where you can canter them into a gateway with nothing in it whatsoever, kick and say 'hup' and they will jump three foot six or three foot nine. Keep to plain wire at this stage.

Whether you go on to school him over barbed wire at home, or leave it until you hunt to jump barbed wire, is purely a matter of choice. I school over plain wire at home, and then when I am hunting, if I come to barbed wire with an experienced horse I just canter up to it, kick him 'hup' and go straight over it. With a green horse I trot up to the wire, let him feel it, go back and canter into it, kick him 'hup' and over. I can say quite honestly, that other than two scratches, I have never had a horse cut jumping barbed wire. And I have only had a horse caught up in barbed wire on two occasions, and on each of these I jumped a five foot fence when hounds were running, and landed straight into a barbed wire fence standing six feet the other side – that is, in both cases I was unaware that I was jumping wire at all. This is something that is most unlikely to happen now. In those days, about twenty years ago, some herds were tuberculin tested and others were not, so a lot of dairy farms had barbed wire fences six foot out from the normal fence, to isolate their cattle from their neighbour's. Today you won't find such double fences, but you will find barbed wire fences, and these will give you no trouble, provided you school your horses properly at home.

One special situation that you may meet with cross country, is one where you jump up onto a bank, and then have to slide down off the bank and jump a fence the other side. The answer here is pure straightforward schooling at home. In our ordinary riding, we teach all our horses to slide down very steep banks, which is easy in this area, because there are a lot of steep slopes. Once we've taught a horse to slide down a bank, so that he is sitting on his hocks, we put a low rail eight or ten feet out from the bank, so he has to slide straight off his hocks, out over the rail. Again we increase the height of the rail, until he is jumping three foot six or four foot straight over without any difficulty at all.

17: *Fanny*

I first saw Fanny as a yearling. She belonged to a friend, who was selling her at the horse sale in Llanybyther. Unfortunately I was particularly hard up at the time and I couldn't buy her, but as luck would have it he didn't sell her. The following spring, I was over at his place, and I saw her again. She had grown very big and very strong, and extremely wild. He wanted to run his cob stallion with his mares, and he didn't want Fanny to produce a foal, so he asked me if I would take her for the summer, and just mess about with her generally. I said yes – he was lending me three other horses for pony trekking anyway, so I didn't have much choice, and Fanny came over three or four days later with the other horses. Her owner, when he came out of the lorry, was pouring with sweat and in a flaming temper. 'I've spent a whole day racing round a field, trying to catch that something-something-something bitch!' He'd got her partitioned off from the other three in the front of the lorry, so we let the other three out into the field, and drove the lorry back to the yard. I climbed over the partition to where Fanny was breathing fire at all and sundry.

I stood quietly for five or ten minutes, and slowly but surely she relaxed, and I put my hand on her. She flinched as though she had been touched with a red hot iron, and cowered back against the front of the lorry. Again I waited for her to relax, and again I put my hand on her. She flinched this time, but not as badly, and, slowly and gently, I caressed her side, just beside her shoulder, working my hand up her side until I got to her back and withers, then I started squeezing and scratching her mane. This she obviously enjoyed and relaxed visibly. I scratched her mane until I got to the back of her ears, and then I went back down again to her side. Then I went away, to collect an old rope halter, came back and put my hand on her

191

again. Again she flinched, but she was getting used to me and allowed me slowly to work the halter over her, up her side and over to the other side of her body, and then to work my hand up her neck, so that her head was between my body and the halter on the other side.

When I got as far as her head, I eased her nose through the halter. Scratching the side of her face, I brought the halter up and over her ears.

Having got the halter on, I attached it to a long plough line, so that we had plenty of rope to play with, undid the partition and swung it back. Then I went back to her and, just rubbing the side of her face, got a hand on the halter, and pulled her round quite easily until she was facing the back of the lorry. As soon as she saw freedom in front of her, she gave a bound forward, tearing the halter out of my hand. This I had foreseen, having twisted the line round the gate post, so that as she bounded out of the lorry, preparing to go hell for leather down the road, she came to the end of the rope within five yards and had to stop. She fought the halter for two or three minutes, then I walked up to her, talking quietly and gently. She sidled away from me, but she couldn't go back because of the halter rope, she couldn't go forward because my friend was standing there, and she couldn't go sideways because there was a bank in the way so she stood, and let me get my hand on her. Again, I went up her side and her neck and got my hand on the halter. My wife got behind her, and Fanny walked forward with my hand on the halter, just holding it loosely, while my friend undid the turn on the gatepost, and took a turn round a tree in the side of the yard instead. We progressed like this until we got to the stable door. Fanny's owner went in through the door, and took a twist round the post in the corner of the stable. Fanny looked in at the stable, looked left and right, and found it uninviting, so she hung back. I stepped back two or three paces and put my hand on her bottom. As soon as I touched her bottom she jumped forward, taking her half into the stable. I walked forwards, slapped her bottom, and in she went like a bullet. We shut the door and I put a feed, some water and hay in for her.

We left Fanny where she was for the night, and the following morning, when I went in to feed her, I found that she hadn't touched the corn, but had drunk the water and eaten the hay. Obviously she had no idea what horse and pony nuts were, and had to be taught to eat them. That was an easy lesson. All I had to do was get my daughter Paddy's old pony, Strawberry, put him in the stable and leave him ten minutes. Of course, as soon as he saw the nuts, Strawberry jammed his head in the bucket and started eating, as if it was the last meal he was ever likely to see.

I stood back a fair distance and watched Fanny, and Fanny watched Strawberry, and after a few minutes Fanny realized that whatever was in the bucket must be good to eat, so she walked over to investigate. Strawberry swung his backside over at her and told her to push off, but Fanny wasn't having any. At this sign of bad manners on Strawberry's part, she bit him hard, to show him that this was no way to treat a lady. Strawberry swung round, but didn't take his head out of the bucket, so Fanny bit him harder, this time catching hold of his ear. This brought his head up quickly, whereupon Fanny seized the opportunity and put her head in the bucket, to see what it was that Strawberry was fighting to protect.

She took a mouthful of four or five nuts, picked her head up and chewed them slowly, whereupon Strawberry put his head back into the bucket, and nothing she could do would shift him again. When she had chewed the nuts and decided that they tasted quite nice, she tried to push Strawberry away, but this time he snapped at her, and eventually, when after five or ten minutes he had finished her feed, I led him out and put another feed in for Fanny.

Fanny walked over this time and sniffed, nibbled a few and then nibbled some more, then really got down and cleared the whole lot. This was an important lesson : once she had learned that nuts were good to eat, training and handling her would be comparatively easy, because nuts could be used as a reward. When she did what I told her, she got nuts – when she didn't, she got smacked.

Three or four times a day I went into her stable, first of all stroking her and handling her. Then I took the stable rubber and scratched her in her ticklish places, untangling a hair here and a hair there, not too much at a time because untangling a tangled mane tends to be a little bit painful for the horse. Over the course of the week I got her mane reasonably tidy and the worst of the tangles out of her tail. By this time, I'd also got her used to being brushed by the body brush, and she quite enjoyed having the particularly itchy bits along her back scratched with the body brush as well.

At the end of the week, I walked in and, instead of a halter, I took the snaffle bridle with me. In my hand I had a few nuts, so she stuffed her nose into my hand to get the nuts, and it was quite simple to get my thumb into the corner of her mouth, so that as she opened her mouth I slipped the bit in and put the rest of the bridle straight over her ears and nose.

I then took the saddle, which I had already placed on the half door of the stable, and slapped it on her back as if it was a normal, everyday thing. I'd put a bucket of nuts in the corner of the stable, before I put the saddle on, so though she sidled away slightly, she went on eating. I reached under her, caught hold of the girth, and gently pulled it under her, then buckled it loosely. By degrees I tightened the girth. She didn't move or flinch in any way.

My wife gave me a leg up until I was lying across Fanny's back. This I hadn't been able to do when I was grooming her, because she was a good 15.3 and all I had been able to do was to reach across her slightly when I was handling her. But even though this was the first time I'd put my weight on her she didn't shift her head out of the bucket. Slowly and gently I rubbed her side and back, and then I brought my leg over the saddle, as far as possible making sure that I didn't touch her side with my knee or my foot as I did so.

Eventually I was sitting in the saddle with my feet one either side of her. I put one foot in one stirrup whilst my wife put the other foot into the other, and just sat there talking to

Fanny whilst she finished her feed. Then my wife opened the door and led her out of the yard and up the road.

We hadn't gone ten or fifteen yards, before I could feel that the mare was completely and utterly relaxed, and I had complete empathy with her. I said 'whoa', just touching the reins. Fanny stopped. In fact throughout the whole of her early gentling, she was one of the easiest horses I have ever dealt with.

We walked about half a mile and the mare was going so well that I risked something that I've only done four or five times in my life. Normally we ride up the hill and when we get to the top I get off and lead the horse back down, because if something is going to go wrong, it is most likely to go wrong going down hill and towards home. But in this case the mare was so relaxed and going so well, that when we got to the top of the hill, and I said 'whoa' and Fanny stopped, I simply twisted my whole body round so that my knee was pressing into the saddle on the near side, and my leg was slightly back on the off side. I took the reins over, so that the near rein was bearing on the left side of the neck, and the off rein out at a wide angle to the right. Then I turned to look down the hill and Fanny, without any hesitation, turned round and walked down.

I could hardly believe it. I had never had a horse that learned so easily and responded so well. When we had gone twenty yards I neck reined her from the left hand side of the road, to the right hand side of the road, then back and forth four or five times. I stopped every twenty or thirty yards, and then we went on again, neck reining and stopping, back into the yard.

We got to the stable door and I was going to get off, when I suddenly wondered whether, since the mare had already stopped, neck reined, turned, she would also back. So I gently tightened the reins so that her head came into her chest, squeezed her with my heels and she took a stride backwards. Leslie caught hold of her head, as I eased myself out of the saddle and on to the ground. I led her into the stable, took her saddle and bridle off, and made a terrific fuss of her. Leslie brought her feed and put it in. She saw the bucket put on the ground, brought her head back and knocked me to one side, telling me to get out of the way, because I was interfering between her and her dinner. I scratched her neck and told her how clever she was, and left her to settle down quietly.

The following day I had her out again, and half way up the hill I clicked my tongue and squeezed her with my legs and she broke into a long smooth striding trot. We trotted five or ten yards, before I pulled her back to a walk, we walked, and I trotted her again. While she was trotting I neck reined her from one side of the road to the other.

The response I was getting from the mare was beyond description. The following day we got old Cork Beg out, and Leslie rode Cork Beg while I rode Fanny. We went for a ride in the forest in a big circle, walking, trotting, neck reining, stopping, backing. I hadn't intended to do more than this, but I hadn't taken into my calculations old Cork Beg.

We came to one of the spots where Cork Beg enjoyed a gallop; we had already passed two of them and he'd behaved himself, but this time he was having fun. He heaved the reins out of Leslie's hands and he was off up the slope, like a bat out of hell. Fanny, who had been trotting gently behind him, suddenly saw Cork Beg disappearing into the distance and this was too much for her. She gave a little squeal and a very small buck, and was straight off after him. Cork Beg steadied at the top of the hill and she steadied back to a trot twenty yards before she got to him, partly because I steadied her back, and partly because she was getting tired. I could feel that she was getting blown, and it was an excellent opportunity for me to assert control. We rode on slowly and gently for another half mile, until we came to a little lane, where we had half a dozen two-foot fences forty or fifty yards apart. Leslie said 'You're never going to jump the mare?' I said 'I've never done this in my life before, but I must see whether it's possible.' Cork Beg, who had been dancing around, half bucking, tearing at his bit, heard what I said, and was gone like a shot from a gun, straight up the lane and over the fences. Fanny beetled after him, determined that she was going to miss none of the fun.

She came to the first of the little brush fences, cocked her large ears, looked at it, steadied and I thought, 'this is it, she's gong to stop', but not a bit of it. She steadied herself at the exact spot, launched herself into mid air, and jumped at least five feet. I jumped at least seven feet, and came back down into the saddle all arms and legs. Fanny should, by all the rules of everything that is natural, have taken the opportunity offered her, and bucked me off there and then, but she didn't. She was away into the next fence after Cork Beg, again steadying and jumping perfectly, although this time she only jumped about a foot higher than the fence, and after that she jumped each fence without any difficulty at all, until she came to the last one. This was a fairish tree trunk lying across the lane. It had no ground line, which foxed her a bit, and she rapped it hard with her front legs. Undismayed, she twisted her hind legs under, and sorted her front legs out without any difficulty. I steadied her back to a trot and we trotted up the road to where the old man, Cork Beg, was waiting for us.

Fanny had had enough, and after petting and talking, and making a fuss of her, I rode her home. I gave her owner a ring that night to tell him how well she was doing, and also that her feet were in a bad state – she had sand crack and she was beginning to get a bit sore – so would he come and shoe her? He does all his own shoeing, and I hadn't finished milking the following morning when he was on the doorstep. He walked into the stable, put a feed down for Fanny, and had a set of shoes tapped on to all four feet within a quarter of an hour, before Fanny had finished her breakfast. Other than moving a bit while he was clenching the nails down, she took no notice of being shod whatsoever.

I rode her for another two weeks, keeping the jumping down to about two foot six. I know that it is supposed to be wrong to jump a two-year-old, but on the other hand Fanny was enjoying it, and the point in training she'd got to in the last two weeks was where I would normally expect a horse to be after six months. She was stopping, turning, neck reining, changing legs at a canter, jumping two foot six and going absolutely superbly. The only thing that she was doing wrong was that once I'd got her eating plenty of corn and a little bit fit, she would dance instead of walk. This was her expression of her keenness, and try as I would I could never really cure her of it. For the following three years, I worked and worked on this point, but except when I was getting her fit for racing as a five-and six-year-old, when she did settle down and walk slowly and steadily, she always tended to dance, mainly because she was enjoying herself so much. She was so full of enthusiasm that her excitement got the better of her discipline, and she would juggle along, wanting to get on with it, to gallop or jump.

After another fortnight, the owner came over again and we had a long consultation about the work we wanted to do with Fanny. I'd already got far further than I'd thought possible. Now she needed shoeing again and the consultation was about whether to take her shoes off and leave her, or to shoe her again and eventually we came to the decision that to get her feet right again it would be best to leave a light set of shoes on her, changing them every three weeks, and turn her out for the summer. So we turned her out and left her. By turning her out and leaving her, of course, I don't mean that I turned her out on the far end of the farm and forgot all about her. I mean that I put her in the field next to the yard, and I wandered out with a small feed for her every day. She spent a lot of time standing up by the yard, and I'd chat to her as I went by, and she'd watch what the other horses were doing, and learn how they behaved.

She went back to her owner for the winter and came to me the following spring, as a three-year-old. During that summer I worked her for a couple of hours, three or four times a week, occasionally popping over little fences, but generally developing her muscles, schooling her and getting her going sensibly and well. Then at the beginning of August, a friend of mine held a hunter trial. We had a lot of horses going and I decided to take Fanny as well, to see how she'd do, so ten days before the hunter trial we put up a set of fences, about two foot six, timber mostly, some brush and a couple of combinations. I jumped her over the first timber fence, intending to do the three smaller fences, but not a bit of it! She was over the first fence, tearing to get into the second, and she went right round the course, giving me very little choice about it – although if the mare wanted to go, I was the last person in the world to stop her. Over the next week we raised the fences until they were three foot six and four foot spreads. The spreads and the doubles were regulation size as well.

We took her up to the hunter trial and rode two other horses, including little Dart, who was there by accident – I just took him round to see how he would do. To my amazement I finished first and third, in the novice hunter trial – first on Fanny and third on Dart.

The thing that finally crowned the afternoon was the fact that my daughter Paddy, who was then ten years old, was put up on an old racing pony that we had, and I took her with me, riding Cork Beg myself, in the pairs. Since the course was in a convenient circle, we went round the pairs course absolutely flat out, steeplechasing stride for stride – neither the old mare nor the old man was going to let the other in front – and we finished third. So we had a wonderful day, and I had suddenly discovered that I had two brilliant young horses in the stable, to go campaigning with.

We took Fanny to another hunter trial in the autumn, and she was third in the open. I think I could probably have won it on her, but since the fences were very big and very strong, I took my time over them, instead of letting her battle on as she would have preferred to do, and we were beaten on time.

Again she went back to her owner for the winter, and came back to me in the spring. This time I really got down to work on her, so that she would jump, and change legs, and I reached the stage where I was purposely putting her into big fences and letting her get herself out of trouble, in this way making her an extremely good hunter, racehorse and hunter trial horse. We had her in three hunter trials, and she was unbeaten in all three of them, because by now her fencing was so good that I was just setting her alight at the start, and I could stop, check, turn and go where necessary.

I kept her that autumn and had her hunting about a dozen times, stopping just before Christmas to get her ready for point to pointing. By the end of February she was extremely fit, and I took her to the first point-to-point. She wasn't really big enough or well bred enough to make a racehorse, but I had high hopes for her because she seemed to have a fairish turn of foot, and she stayed forever. We had a very good point-to-point season with her, she was third or fourth in all of her point-to-points until the last one, which she won with ease. So we turned her away for the summer to have a good rest.

At last I had got the horse I was looking for. I had one object and one only in mind, and that was the Foxhunter's Chase at Aintree. I had always wanted to ride at Aintree, but until Fanny I'd never had a horse that I thought capable of tackling the Aintree fences. That autumn we worked on her, and in due course we had her fit and qualified. All the paper work, registrations, hunter's certificates, jockey's certificates and training permits were in order by the end of February. We took her up to Leicester for the first race. The going was very, very heavy, and it was a two and a half mile race. They went away like bats out of hell and the pace was too fast for Fanny, so I settled down well in the rear of the field for the first circuit, and the other horses started coming back to me. Three fences from home I was lying fifth in a field of thirteen. We came over the open ditch, and there was a horse lying on the ground in front of us. Fanny made a valiant effort to jump the horse as well as the fence, but the mud was just too much for us and over we went. It was the only time she ever fell racing in her life, but I was more than satisfied with her performance.

We took her to Bangor on Dee a fortnight later, and the going was very fast, and she just hadn't got the pace to keep up with them. She jumped impeccably and went well, so I went home and decided that all I could do after that was to pray hard every night that it would rain hard every night for the next fortnight. I must have offended the Almighty in some way, because it didn't, it stayed as dry as a bone. The Foxhunter's Chase at Aintree was on the Thursday and we spent the whole of the Monday, Tuesday and Wednesday getting the landrover greased and oiled – an unprecedented experience for the landrover; it nearly fell to pieces with the shock. Then we cleaned the trailer out, my wife repainted it, and Paddy never stopped grooming Fanny, so that by Wednesday we were all worn out.

I got my racing stuff packed into the back of the landrover during Wednesday and at half past three on Thursday morning the alarm went off and we got up, gave Fanny a feed and ourselves a quick cup of coffee. By half past four we were on the road for Aintree. By about half past eight we passed through Hereford, and up over the downs towards Worcester. It was a glorious morning, the birds were singing, and we stopped to get the *Sporting Life* to see our names in print for Liverpool that day. Paddy, who had woken up – she had been fast asleep in her sleeping bag in the passenger seat – proceeded to read out what they'd said about each horse. She came to Fanny: 'Third rate point-to-pointer: has won a small point-to-point, and has been placed in bad company'. But still hope springs eternal in the human breast, and at Aintree all things are possible. The rest of them might fall and we might win.

We drove on and on the endless miles to Liverpool. Eventually we came to a sign which said 'Aintree Racecourse' and every two or three miles we passed another until suddenly the signs to Aintree ended and I decided I'd gone too far. By the side of us were walls thirty feet high. Her Majesty's Prison at Walton might be where I should have been, paying for my misdeeds, but it certainly wasn't where I was supposed to be racing that day! So I turned round and retraced my steps, and finally after some searching found concealed without any sign or indication, a small gate. We went in and there at last I was at the Mecca of steeplechasing.

We got Fanny out and walked her round, and let her pick grass. Her silver mane and tail floated in the breeze looking an absolute picture, but compared with most of the other horses, she looked like a weight carrying hunter. Having attended to her wants, we put her back into the trailer and went to inspect Aintree. In some ways it was a bit shabby, and in others it was magnificent. Fanny ran in Paddy's name and as owner she was due a free lunch and a seat in the county stand, so she went off to her meal, at the expense of Mrs Mirabelle Topham, whilst I went and dumped my racing kit, and greeted my various acquaintances in the jockeys' room.

An old friend, who had a friend who had ridden at Aintree before the war, had given me a piece of invaluable advice. 'Whatever you do', he said, 'do not go and look at the course. If you stand and look at the fences, they look terrible; when you're riding them, they don't look bad at all.' So I carefully avoided the course and went to have a look round. Eventually the crowds arrived and Paddy and I, after some difficulty, eventually located Leslie and Fanny's owner, who had driven up separately. We saw the first race, and then the rest of them went to get Fanny out and ready, whilst I went in and got changed. When I was ready I sat on my bench in the jockeys' room in glum despondency: I'd been a fool to think that I could take a fairish hunter up to Aintree, and expect a poor little mare of 15.3 to jump the biggest and most severe fences in England.

But before long the call came, 'jockeys out', and we trooped out to see Fanny swagger around with her head in the air, and her silver tail floating behind her. The television cameras seemed to be focusing on her, and she seemed to be cavorting to show herself off, for the benefit of a million viewers.

'Jockeys up'. Fanny came in and I tightened her girth. Her owner gave me a leg up, and we walked round the paddock, then they filed out of the paddock and, by a bit of skulduggery, I indulged my favourite superstition, based on the old-fashioned saying 'Last out, first in'. I always find that if I don't observe this rule, I go for a burton. So I jumped off Fanny and pretended to readjust the breast plate, so that I was last out of the paddock, and we cantered down to the start.

The starter called the roll and we were walking around. He said 'Come into line' and, intending to be clever, I said 'No, no, no, I'm not ready yet,' but I couldn't catch him out, he was too old and wily a bird. So up went the tapes and away we went, but instead of getting a flying start as I intended, I lost a couple of lengths and was at the tail of the field.

Fanny was reaching for the reins and galloped into the first fence, going like a bat out of hell. She stood back and flew it like a bird, and so with the second and the third. Then came the Chair; this is supposed to be a huge fence. I looked at it afterwards and it did look huge; two of us stood on either side of it and we couldn't see each other, while the ditch was wide enough to drive a landrover down. But as we galloped into it, it looked a nice, inviting little fence. Fanny stood back and I don't think she touched a twig.

We went into the water jump. Fanny had jumped water jumps before, and knew what they were about. Unfortunately what I had forgotten to tell her, was that at Aintree it's not twelve feet of water, but fourteen feet, and she dropped her hind legs into it. We lost a length or two, but she heaved herself out and away we went in pursuit of the rest of the field, across the Melling road, and into what is the first fence of the Grand National. Each fence she measured perfectly, and each fence she jumped without touching it. We came to the open ditch before Beechers and, much to my amusement, there was a jockey sitting on top of it, looking down at his horse lying on the ground below. Fanny stood right back and skimmed it like a hurdle, and so into Beechers. She jumped Beechers and in mid air, I looked down and saw a horse lying on the ground, just where we were going to land. Fanny checked and landed short of the horse, took off and jumped him. There was another horse straight in her path, she changed legs, nearly throwing me from the saddle, and went round him and on into the next fence.

Going towards the Canal turn, I was beginning to see bodies everywhere, and instead of the sixteen starters, there were only six in front of us. She came round the Canal turn on one leg, and just popped over the corner, almost from a standstill, down the far side. I was feeling wonderful. All the difficult fences were over and done with, and we just had five or six fences to hurdle, and that was it. Each fence was about fifty yards wide and though they were big and strong, their width made them look small and inviting. We came into yet another seemingly easy fence. Fanny took off well back, and I looked down and saw what appeared to be a dirty great river beneath us; I'd forgotten that we still had Valentine's Brook to jump, and this was it. We seemed to stay suspended in mid air forever but eventually down we came, with Fanny swinging away, as if she'd popped over one of the little schooling fences at home.

We jumped the next fence, and there lying on the ground was another horse – we were sixth. Over the last three fences as if they weren't there, and up round the elbow and up the straight to the winning post. Admittedly we'd been last from the very beginning to the end, but the fact remained that I had completed a life's ambition, I had ridden round Aintree. And because I was riding one of the very best fencers in the country, I had found it extremely exhilarating and easy, and was walking on cloud nine.

We rode back to the paddock and I unsaddled. I went back to the weighing room feeling absolutely elated. So obvious was my joy and happiness, that Lord Leverhulme, who happened to be speaking to one of the other stewards, stopped me and said 'There's a picture of a very happy man'. There was only one reply; I said 'You're looking at the happiest man in England; I've just ridden the best horse in England, over the best fences in England.' He said 'Did you win?' and I said 'No, I came tailing round at the back of the field, but she is still the best horse in England, and those are still the most fantastic fences in England'.

I changed and went back to the rest of the family. We loaded Fanny, watched a couple of hurdle races and flat races, and went home. We got to Tregarron at about two o'clock in the morning, and I was almost asleep. Paddy, who was supposed to be staying awake to talk to me, would talk hard for two minutes, and then go back to sleep herself.

I drove on until I came to one of my local pubs, which being out in the country has no idea of licensing hours. By now it was nearly four o'clock and I thought, 'I'm so near asleep

that if I don't get someone to drive me home, I'm going to have
an accident'. So I got out and went round the back of the pub,
to see if I could get a chauffeur. I was just going to go in, when
it suddenly struck me that if they had been in there drinking
since closing time, they were even more likely to have an
accident than I was. So I went back to the landrover and drove
the last ten miles home, getting back at half past four. I un-
loaded Fanny and put her into bed, which Leslie had got
ready with food and water. Fanny walked over to the bucket,
had a drink, went over to her feed bucket and tucked into it,
as if it was the most normal thing in the world to be driven
round the countryside for twenty-four hours and raced around
Aintree. I had a quick cup of coffee and went to bed, and slept
solidly until half past two that afternoon. Leslie who had got
up early and done the horses and all the work, said that
when she'd gone out Fanny was fast asleep, until she'd brought
another bucket of water and a feed, whereupon Fanny had
got up, had a drink, eaten her breakfast and laid down
again.

I went out to see how she was, and there she was lying fast
asleep on the floor. I felt her legs which were cool. She opened her
eyes and looked at me, and I patted her shoulder and scratched
her neck. I sat down in the straw between her legs, leaned back
against her tummy and we talked to each other for the next half
hour, about the superb time we'd had the day before. After
that, for the rest of the season, Paddy, who had just passed her
eighteenth birthday, rode her in Ladies' races, never being out
of the first three, and I went back to riding and schooling young
horses.

Eventually I managed to scrape enough money together to
buy Fanny, intending to ride her round Aintree again, but unfortunately in one of her schooling races, the ground was very
rough, and she put her foot in a rut, cracking the fetlock bone
very slightly. It was enough to put her out for the next nine
months, and we never got to Aintree again.

My racing days are over too. I had an accident a year ago
and injured my spine, so that Fanny and I will not face any more

fences together. Yet I had her nine years, competing and hunting, and I don't think I ever rode a better horse.

I remember on one occasion, when I was out hunting with
her with the South Pembrokeshire, we breezed into a bank
six foot high and six foot wide. I expected her to bank it, but
not a bit of it, that wasn't Fanny's fashion. She accelerated
and flew it, clearing the whole damn lot, and I found an
eight foot drop on the other side. She came down, landed,
changed legs and went straight away. She jumped barbed wire
the way most horses jump hurdles. Nine times out of ten, when
you wanted to catch her, she would come over, and almost put
the halter on herself. But sometimes she would choose to trot
round you in an infuriating manner, in a circle fifteen yards
across, with you in the middle. You could see her roaring with
laughter and saying 'Now, you stupid beggar, catch me if you
can'. I would give up in disgust, go up and open the gate.
Then she would trot up the road behind me, with her head on
my shoulder.

She's gone now to the stallion, and we hope that she's going
to produce something with a bit more pace and size than she
has, and if she does and he has only half as much heart, and
half as much jumping ability as his mother, he's going to be
a very, very good horse indeed, and sometimes I think that even
if I'm still in a wheelchair, I'll get on that horse, and I'll ride
him round Aintree, come hell or high water, even if I die in the
attempt.